THE NEW SCIENCE OF ORGANIZATIONS

THE NEW SCIENCE OF ORGANIZATIONS
A Reconceptualization of the Wealth of Nations

Alberto Guerreiro Ramos

UNIVERSITY OF TORONTO PRESS
Toronto Buffalo London

© University of Toronto Press 1981
Toronto Buffalo London
Printed in Canada

ISBN 0-8020-5527-3

HM
201
·R35

Canadian Cataloguing in Publication Data

Ramos, Alberto Guerreiro, 1915–
    The new science of organizations
    Includes index.
    ISBN 0-8020-5527-3
    1. Social systems. 2. Organization. I. Title.
    HM201.R35   303   C81-094130-9

Publication of this book has been assisted by the Federal University of Santa Catarina, Florianopolis, Brazil, and the Publications Fund of the University of Toronto Press.

To my wife Clelia,

my children, Eliana and Alberto,

and my grandchildren,

Tatiana, Chara, Leah, Allison, Andrew Victor,

and Henrique Alberto.

'Nothing is more reprehensible than to derive the laws prescribing what *ought to be done*, from what *is done*, or to impose upon them the limits by which the latter is circumscribed.'

Kant, *Critique of Pure Reason*

'Most of us have known adolescents who were compulsive eaters. The compulsive eater, stuffing himself so urgently and inattentively that he scarcely notices what he devours, imposes upon himself two quite serious penalties. Though vast quantities of the truly "good" are being crammed into his body, he is all the while becoming less and less fit. And he is, alas, not enjoying what he eats; he scarcely remembers to taste it.

As twentieth-century adults, we have become compulsive users, with approximately the same results.'

Walter Kerr, *The Decline of Pleasure*

Contents

Preface

In this book I set forth the conceptual framework for a new science of organizations. My objective is to contrast a multicentered model of social systems analysis and organizational design with the current market-centered model which has dominated private enterprises and public agencies for the last eighty years. Generally I argue that a market-centered theory of organizations is applicable not to all, but only to a special type of activity. Attempts to apply its principles to all forms of activity are hindering the actualization of possible new social systems needed to overcome the basic dilemmas of our society. I further contend that the model of manpower and resource allocation prescribed by this dominant organization theory is not mindful of ecological requirements and is therefore not commensurate with the potentialities of the contemporary state of productive capabilities. Finally, I suggest that the manner in which the dominant model is taught is deceptive and has disastrous consequences since the limits of its functional character are not acknowledged.

I am using the expression 'new science of organizations' in a broad sense. It includes not only matters pertinent to fields currently labelled as business and public administration, but also themes specifically pertinent to economics, political science, policy science, and to social science in general. Thus, as conceived in this book, the new science of organizations addresses itself to problems of ordering social and personal affairs in a micro as well as a macro perspective.

The socialization process is formally and informally organized in to-day's market-centered nations to make the human individual essentially sensitive to economic inducements. This condition is an episode of psychic deformation from the standpoint of the view of man pervading the argument of this book. To spell out this view is a self-defeating endeavor because any such attempt constitutes a closure upon the potential of human experience. The understanding of man, however, can be continuously drawn from the reading of deeds and texts where wisdom speaks to all human beings irrespective of their historical and racial circumstances. The new science of organizations is not concerned only with social systems required to enable individuals to succeed in a market-centered society, but with issues and problems of social systems design common to all human societies.

By and large the teaching and training offered to students not only in schools of public administration and business, but in departments of social science as well, is still predicated on the assumptions of the market-centered society. An alternative model of thought, not yet articulated in systematic terms, is needed today, because the market-centered society, more than two hundred years after its rise, is now disclosing its limitations and its distortive impingement upon human existence at large. It is no less than such a mode of thought that this book seeks to articulate.

In chapter 1 I deal with the root concept of all social science – reason. Modern social science cannot be adequately explained apart from the peculiar understanding of reason it implies. In this century the critique of modern reason was begun by Max Weber and Karl Mannheim who nevertheless failed to face consistently its complexity. More recently Eric Voegelin has tried to evaluate modern reason from the standpoint of the classical legacy of thought. Significant as his great contribution must be considered, it, however, assumes a restorative character not sufficiently qualified. Moreover, it fails to provide the new science of organizations and society with an operational and analytical expertise, required by the unprecedented historical conditions of our time. Another significant strand of criticism is represented by the so-called Frankfurt school which, I will show, is still fraught with modern historicist fallacies.

Chapter 2 is a critique of the contemporary model of social science from the standpoint of an alternative model which I call a substantive theory of human associated life. It draws upon Max Weber's distinction between *Wertrationalität* (value or substantive rationality) and *Zweckrationalität* (functional rationality) and Karl Polanyi's analysis of the market-centered society.

The third chapter conceptualizes the psychological syndrome inherent in the market-centered society and specifies its main features, namely, the fluidity of the self, perspectivism, formalism, and operationalism. It explains that as long as the citizens at large continue to succumb to the organized inducements, pressures, and influences which keep such a syndrome operative, there will exist, at best, little room for a revitalizing social transformation.

In the next chapter I argue that organization theory, as a disciplinary field, is losing sense of its specific objectives, by attempting to assimilate models and concepts extraneous to its proper domain. In substantiating this argument, I discuss instances of 'misplacement' of concepts exemplified by current issues in the field of organization theory. I conclude chapter 4 by articulating some basic permanent issues of the scientific study of formal organizations.

Chapter 5 presents the concept of cognitive politics. I show that cognitive politics is the most important hidden dimension of the psychology of the market-centered society. Organization theory has never reached the status of a scientific discipline because its proponents are unaware of such a dimension. Standard organization theory is pre-analytical in the sense that it accepts the state of human affairs in the market-centered society as a given, being unaware of a larger range of objective possibilities. The chapter focuses at length upon three unarticulated assumptions of current organization theory, namely, the identification of human nature with the behavioral syndrome inherent in the market-centered society, the definition of the person as a jobholder, and the identification of human communication with instrumental communication.

The sixth chapter lays bare the epistemological blindspots of extant organization theory and concludes by identifying the need for a new science of organizations based on the concept of social systems delimitation. My main critical contentions in this chapter are that current organization theory fails to distinguish systematically between substantive and instrumental rationality, and between the substantive and the formal meaning of the organization; it lacks a clear understanding of the role of symbolic interaction in interpersonal relations at large; and it relies on a mechanomorphic view of man's productive activity.

In chapter 7 I present a multicentric model of social systems analysis and organizational design which I call social systems delimitation. It depicts the market as a necessary, but limited social system, which is appropriately an enclave of society. It conceptualizes other social systems, as for instance isonomy and phenonomy, and describes essential dimensions of the multiple enclaves which are constitutive of the overall social fabric. This chapter is an introductory statement to what I call the paraeconomic paradigm.

The law of requisite adequacy is presented in chapter 8 as a fundamental topic of the new science of organizations. According to this law a variety of social systems constitutes an essential qualification of any society, which is to be responsive to its members' basic needs of production and actualization. Moreover in this chapter I contend that each of these social systems prescribes design requisites of its own. The law of requisite adequacy is illustrated through an analysis of the technology, size, cognition, space, and time requisites of social systems.

In chapter 9 I discuss the policy implications of the paraeconomic paradigm. I contend that social systems delimitation is not only a theory bearing upon the micro-organizational level, but is also applicable at the macro-societal level. I illustrate this contention by discussing the processes of manpower and resource allocation from a delimitative perspective.

I conclude the book with a summary of the main tenets of the new science of organizations and trace the general direction of a research agenda which such a science implies.

The ten chapters of this book constitute an organic unit and must be read in the sequence they are presented, otherwise the reader will miss fundamental aspects of its conceptual thread. This is particularly true of the last chapter, where my policy and philosophical thrusts become evident. It cannot be understood if read as a discrete piece.

Chapters 4, 6, and 7 re-elaborate ideas and themes treated in articles which I had written earlier and were published in *Public Administration Review* and *Administration and Society*, as well as in Carl J. Bellone, ed., *Organization Theory and the New Public Administration* (Allyn and Bacon 1980). I appreciate the permission of these publishers for the use of these materials.

The search for a new science of organizations has been underway for some time. It has been a piecemeal endeavor undertaken painstakingly by a great number of scholars. This book takes advantage of much of their creative effort, and begins to mold it into an encompassing body of knowledge.

Acknowledgments

I want to express my gratitude to the School of Public Administration of the University of Southern California where I have found friendship, intellectual stimulation, and material support to undertake the research from which this book has grown. It is impossible to list all the names of those to whom I am indebted. I must mention first, however, Frank P. Sherwood without whose assistance I could hardly have overcome the difficulties of my transition from Brazil to academia in the United States. He has been a friend, a severe but honest critic of my writings, and sometimes a sponsor of my academic pursuits. I owe a special debt to Henry Reining, David Mars, and E.K. Nelson, who as deans of the USC School of Public Administration in successive periods played a major role in giving me the kind of encouragement I needed in different vicissitudes of my career in this country. I want also to express my appreciation to William K. Leffland who, while associate dean of the school, found a variety of means to facilitate my research work. Significant contribution to the clarification and improvement of my thinking has resulted from criticism and suggestions offered to me by colleagues and other persons who have read either parts or the entire draft of this book. Among them I am glad to mention Peter von Sievers, Ross Clayton, Robert Berkov, Wesley Bjur, Alex McEachern, Karl Scheibe, David L. Schaefer, Daniel Guerrière, Ellis Sandoz, William Dunn, Louis Gawthrop, my graduate students in general, and particularly Francis Cooper, Tom Heeney, George Najjar, Jonathan Moyo, and Levi Reeve Zangai, intimate and insightful collaborators of whom I have availed myself in matters of content and style. Special thanks are due to graduate students Charles M. Dennis and B. Terence Harwick whose review and work on the complete text significantly contributed to its force and clarity.

My stay at Wesleyan University and Yale University as Visiting Professor and Visiting Fellow, respectively, on the occasion of my sabbatical leave from USC during the academic year of 1972–73, represented an auspicious opportunity for me to determine the thrust of this book. The screening process of the University of Toronto Press under the guidance of R.I.K. Davidson has been of inestimable value to the articulation of my work. The critiques of the manuscript I received from University of Toronto Press readers contributed sharply and generously to the present contour and substance of the book.

I could also not forget to thank Constance Rodgers, Mary Priem, Artimese Porter, and others on my school's staff who conscientiously and effectively have given me the secretarial support needed to bring the draft of this book to its present state. Beverly Harwick typed the final draft and I am appreciative of her alert diligence. I would also like to register here my tender regard for two lovely creatures, my dogs, Tupi, now gone forever, and Cochese, faithful and joyful guardians of my personal space and companions in my solitary writing hours.

The responsibility for the contents of this book, of course, rests with me.

THE NEW SCIENCE OF ORGANIZATIONS

1

Critique of Modern Reason and Its Bearing upon Organization Theory

Organization theory as it has prevailed is naive because it is predicated on the instrumental rationality inherent in exant Western social science. In fact until now this *naïveté* has been the fundamental reason for its practical success. Nevertheless we must now recognize that its success has been unidimensional, and, as will be shown, has a distortive impact on human associated life. This is not the first time that because of theoretical considerations one has been led to condemn what, in practical social life, has worked. In fact forty years ago Lord Keynes noted that economic development rose out of avarice, usury, and precaution, all of which he despised. Nevertheless he concluded that 'for a little longer' they must continue 'to be our gods,' because 'only they can lead us out of the tunnel of economic necessity.' Within the precarious conditions still expected for a while, Keynes recommended that 'we should pretend to ourselves and to everyone that fair is foul, and foul is fair; for foul is useful, and fair is not' (Keynes 1932: 372).

Like Keynes, some today would also wish to suspend criticism of current organization theory, because although it may be poor in sophistication, it works. To do this, however, one must pretend that *naïveté* is fair, while theoretical sophistication is foul. There comes a time such as ours in which the psychological energy an individual must spend in order to cope with the tensions resulting from this self-deception is of such magnitude that he refuses to be conventionally successful and ceases to comply with the norms by which society legitimizes itself. In such circumstances organization theory as it is now known is less convincing than it has been in the past. Moreover it becomes impractical and inoperative to the extent that it continues to rely on naive assumptions.

The word *'naïveté'* is used here in the Husserlian sense. Husserl acknowledged that the essence of the technological and economic success of advanced industrial societies has been an outcome of intensive application of natural sciences. However, the manipulative capability of these sciences is not necessarily an indication of their theoretical sophistication. Thus, according to Husserl, to the extent that such sciences take as self-evident the prereflective mode of everyday life, they are 'on a level with the rationality of the Egyptian pyramids' (Husserl 1965: 186).

In other words, Western natural sciences do not rest upon a truly analytical mode of thought as they have become caught up within the frame of immediate practical interest. This state of affairs is, perhaps, what Husserl referred to when he said: 'All natural science is naive in regard to its point of departure. The nature that it will investigate is for it simply there' (Husserl 1965: 85). Natural sciences may eventually be forgiven for their naive objectivism because of their productivity. But this permissiveness cannot hold in the social domain, where erroneous epistemological premises become a crypto-political phenomenon – i.e., a disguised normative dimension of the established power configuration.

The present chapter is an attempt to identify the epistemological thrust of established social science, of which current organization theory is a derivative. My main contention is that established social science is also predicated upon an instrumental rationality, which is peculiarly characteristic of the market system. I will conclude by suggesting the topic of the following chapter: a more theoretically sound concept of rationality, to wit, substantive rationality, which provides the ground for an alternative social science in general, and for a new science of organizations in particular.

REASON AS RECKONING OF CONSEQUENCES

In the modern period of Western intellectual history beginning in the seventeenth century and continuing to this day, the previously established meaning of those words which constitute a basic theoretical language has drastically changed in a determinate direction. In the works of men like Bacon and Hobbes writing in the cultural climate of the seventeenth century, it is evident that the meaning of the term 'reason' (as well as those of other terms such as 'science' and 'nature') was already peculiar to an unprecedented semantic universe.

In the age-old sense, as will be shown, reason was understood to be a force active in the human psyche which enables the individual to distinguish between good and evil, false and genuine knowledge, and, accordingly, to order his personal and social life. Moreover the life of reason in the human psyche was envisioned as a reality which resists being reduced to a historical or social phenomenon.

In Hobbes's works 'modern reason' is for the first time clearly and systematically articulated, and its influence has not perished to this day. Defining reason as a capacity which the individual acquires 'by industry' (Hobbes 1974: 45) and which enables him to achieve nothing more than a 'reckoning of consequences' (Hobbes 1974: 41), Hobbes intended to strip reason of any normative role in the domain of theory building and human associated life. In a book in which he attempts to substantiate such an intent, he says, 'Civil philosophy' is 'no older ... than my book *De Cive*' (Hobbes 1839: ix). Mainstream social science in both its academic and popularised versions is largely a footnote to Hobbes.

It seems that the term 'rationality' is usually now employed by laymen as well as by social scientists in a deceitful fashion, but a deceit based no longer on the self-conscious type of inquiry conducted by Hobbes but on a profound bewilderment. The deceitful implications which the term holds at present need to be identified for what they are. Since nowadays rationality very often assumes connotations antithetical to the fundamental aims of human existence, anti-rationality without qualification has become one of the theses of some who see themselves as humanists. When one examines their intentions, however, one realizes that theirs is a mistaken cause. Their intentions may be sound, but their target is deceptively misplaced. The rationality they fight is actually the distortion of a key concept of individual and associated life.

The transvaluation of reason – leading to the conversion of the concrete into the abstract, of the good into the functional, and even of the ethical into the a-ethical – characterizes the intellectual profile of writers who have attempted to legitimize modern society solely upon utilitarian grounds. A major thesis of this book will be to point out that when compared with other societies, modern society has demonstrated a high capability for absorbing, by distortion, words and concepts whose original meaning would clash with its self-sustaining process. As the word reason could hardly be discarded because of its centrality to human life, modern society has made it compatible with its normative framework. Thus in modern market-centered society, predicated upon the Hobbesian understanding of rationality, distorted language has become normal, and one of the means to critique such a society is to describe its cunning in misappropriating the theoretical vocabulary which prevailed before its rise.

With the intent of paving the way toward a new science of organizations and society at large, free from distorted theoretical language, in the following paragraphs I briefly discuss the critical assessement of modern reason undertaken by leading contemporary scholars.

RESIGNATION AND MAX WEBER'S VIEWS ON RATIONALITY

When Max Weber undertook his academic work, the age-old notion of reason had already lost the normative connotation it always had as a referent for ordering personal and social affairs. On the one hand, from Hobbes to Adam Smith and the modern social scientists in general, instincts, passions, interests, and sheer motivation replaced reason as the referent for understanding and ordaining human associated life. On the other hand, under the sway of the Enlightenment, from Turgot to Marx, history replaced man as the carrier of reason. Against this situation Max Weber stands as a solitary figure. He rejected both the crude British empiricism and the naturalism of social scientists as well as the historical determinism mainly characteristic of German thinkers. One clear indicator of the polemical overtone of Weber's academic work is his attempt to qualify the notion of rationality.

Max Weber is frequently portrayed as a true believer in the unqualified excellence of the logic inherent in the market-centered society. A faithful reading of his work suggests, however, a different account of his thought on this matter. He did write extensively on the market as the most effective arrangement for enhancing the productive capabilities of a nation and for escalating its process of capital formation. But in addressing himself to the market and its specific logic, it is evident that Weber does not taint his investigation with fundamentalism. He was not a fundamentalist because he explained the market and its specific logic as the thrust of a singular epoch: history, for Weber, does not end its course with the advent of such an epoch. He focuses upon these matters from the standpoint of functional analysis. In fact, he deserves to be considered as the founder of functional analysis. Modern authors, as for instance Adam Smith, overlook the precarious character of the market logic, while Max Weber interprets it as a functional requirement of a determinate episodical social system. Adam Smith proceeded as a fundamentalist since he exalted the market logic as a normative ethos of human existence in general. Max Weber, however, describes such a logic (of which bureaucracy is a manifestation) as a heuristic construct congenial to a peculiar form of society – capitalism or modern mass society. Max

Weber explicitly condemns any fundamentalist type of economic analysis which 'identifies the psychologically existent with the ethically valid' (Weber 1969: 44). In this vein, having in mind the liberal economists, he remarks: 'The extreme free traders ... conceived of [pure economics] as an adequate picture of "natural" reality, i.e., reality not disturbed by human stupidity; and they proceeded to set it up as a moral imperative – as a valid normative ideal – whereas it is only a convenient type to be used in empirical analysis' (Weber 1969: 44).

Max Weber's account of capitalism and modern mass society was essentially critical, in spite of certain laudatory appearances. He was struck by the way such a society was transvaluating the traditional meaning of rationality, a process that he inwardly regretted, although he failed to confront it directly. Even though Weber refused to build his analysis on moral indignation, as did other theorists, most notably Karl Marx, it is erroneous to attribute to him any dogmatic commitment to the rationality engendered by the capitalist system. Weber's distinction between *Zweckrationalität* and *Wertrationalität* – which he sometimes minimizes – is indicative of his moral conflict with the dominant trends of modern mass society. As is widely known, he pointed out that formal and instrumental rationality (*Zweckrationalität*) is determined by expectation of results or 'calculated ends' (Weber 1968: 24). Substantive or value-rationality (*Wertrationalität*) is determined 'independently of its prospects for success' and does not characterize any human action concerned with 'achievement of a result ulterior to it' (Weber 1968: 24–5). Accordingly Weber describes bureaucracy as exerting rational functions in the peculiar context of a capitalist market-centered society. Its rationality is functional, not substantive. Substantive qualities are an intrinsic component of the human actor.

On no grounds can one consider Max Weber as a representative of bourgeois rationality. He looked at such rationality with clear personal detachment. Those who claim otherwise inadvertently identify his ad hoc remarks with his personal stance in general, and as well fail to realize the spiritual tension underlying his effort to investigate *sine ira ac studio* the thrust of his epoch. To be sure he was unable to solve this tension by undertaking a social analysis from the vantage point of substantive rationality. In fact *Wertrationalität* is, so to speak, only a footnote in his work; it did not play a systematic role in his studies. If it did his research would have taken a completely different path. He chose resignation (i.e., value-neutrality, not confrontation) as a methodological posture in his study of social life. Yet this resignation never bewildered him, turning him into a radical historicist. Significantly he deemed as 'self-deceptive' any position which 'asserts that through the synthesis of several party points of view, or following a line between them, practical norms of *scientific validity* [italics in the original] can be arrived at' (Weber 1969: 58). His historicism was kept in balance by his strong sense of the finitude of scientific concepts as compared with the 'infinitely manifold stream' (Weber 1968: 92) of reality.

Max Weber's qualified functionalism has been misunderstood by some of his interpreters and even some of his self-proclaimed followers. A case in point is Talcott Parsons, whose work Max Weber apparently influenced. Parsons shows little or no moral ambiguity toward the rationality inherent in the market system. In the light of his dogmatic model of structural-functional analysis from which he derives his notion of 'pattern variables' and 'evolutionary-universals,'[1] the specific requirements of advanced capitalist society become dogmatic standards for comparative social science, and even for history itself.

KARL MANNHEIM'S
ARRESTED INSIGHT INTO RATIONALITY

Karl Mannheim obviously relies on Max Weber in order to elaborate a distinction between substantial and functional rationality. He defines substantial rationality as 'an act of thought which reveals intelligent insights into interrelations of events in a given situation' (Mannheim 1940: 53) and suggests that acts of this nature make possible a personal life oriented by 'independent judgements' (Mannheim 1940: 58). This rationality constitutes the ground of ethical, responsible human life. Functional rationality relates to any conduct, event, or object insofar as it is recognized as purely a means to attain a given goal. The unlimited impingement of functional rationality upon human life undermines its ethical qualifications.

This distinction is thus worked out for ethical purposes. Indeed, Mannheim stresses that functional rationality is 'bound to deprive the average individual' (Mannheim 1940: 58) of his capacity for sound judgment. He sees a decline of the critical faculties of the individual proportional to the development of industrialization. He also suggests that although functional rationality has existed in early societies, there it was restricted to limited spheres. In modern society, however, it tends to encompass the whole of human life, leaving to the average individual no choice but to give up his autonomy and 'his own interpretation of events for those which others give him' (Mannheim 1940: 59). His book *Man and Society in an Age of Reconstruction* is an inquiry about how to safeguard human life against the increasing expansion of functional rationality. Mannheim claims that whoever wants to be consistent with the distinction between the two kinds of rationality must realize that a high degree of technical and economic development can be coincident with low ethical development. This point is worth underlining because there are authors of positivist orientation who seem to acknowledge the validity of the distinction, apparently without realizing its ethical consequences.

Mannheim's distinction does not imply that functional rationality is to be abolished from the social realm. It rather stipulates that a true, sound social order does not obtain when the average man loses the psychological strength to stand the tension between functional and substantial rationality and totally surrenders to the claims of the former. This situation is aggravated when students of decision-making rule out in their writings the tension between the two rationalities. By focusing on decision making from a purely technical and pragmatic viewpoint, they accept functional rationality as the primary standard of human life.

Apparently the analysis undertaken by Karl Mannheim assumed a confrontive stance in the sense that it reflects the author's libertarian urge to find means to change the present state of industrial societies. In fact he did not fully draw the consequences of his distinction. His 'relationalist' eclecticism, by which he intended to integrate all the main currents of contemporary social science, ultimately led him to bewilderment. Neither did Mannheim's concern for human freedom save him from intellectual perplexity. His classificatory endeavor of assessing and commenting upon findings in the domain of conventional social science never really allowed him to arrive at a coherent set of theoretical guidelines. For instance, in *Man and Society* sharp analysis and accurate remarks are presented, but ultimately he failed to develop an idea of social science attuned to his notion of substantial rationality.

THE FRANKFURT SCHOOL'S CRITICAL THEORY

Rationality has been one of the central concerns of the so-called Frankfurt school.[2] Its main representatives, in essence, say that in modern society rationality has become a disguised tool for the perpetuation of social repression rather than being a synonym of true reason. These authors intend to restore the role of reason as an ethical category and, therefore, as the referent for a 'critical theory' of society. Apparently they refuse

Marx's assumption that rationality is inherent in history, and that the process of modern society, by dialectically criticizing itself, will lead to the Age of Reason. They point out that what Marx did not realize is that in modern society the productive forces have gained their own independent institutional momentum, thus subordinating the entire human life to goals that have nothing to do with 'human emancipation'.

Horkheimer's and Adorno's questioning of Marx's notion of reason is a logical consequence of their analysis of the Enlightenment tradition. They see the Enlightenment as the moment in which the understanding of reason was severed from its classical legacy. According to Horkheimer, there is a theory of objective reason running from Plato and Aristotle through the scholastics, and even through German Idealism (Horkheimer 1947: 41) which emphasizes ends rather than means, and the ethical implications of the life of reason for human existence. Such a theory 'did not focus on the coordination of behavior and aim, but on concepts – however mythological they sound to us today – on the idea of the greatest good, on the problem of human destiny, and on the way of realization of ultimate goals' (Horkheimer 1946: 5). Horkheimer sees implicit in such an understanding of reason precepts for ordering human existence.

However, the Enlightenment transforms thought into mathematics, qualities into functions, concepts into formulas, and truth into statistical frequencies of averages. In other words, with the Enlightenment 'thought becomes mere tautology' (Horkheimer and Adorno 1972: 27) and reason one's 'adjustive faculty' (Horkheimer 1947: 97).[3] In the perspective of the Enlightenment the world is written in mathematical formulas, and the unknown loses its classical transcendent meaning, and becomes something relative to the available calculative capabilities. Thus Horkheimer and Adorno write: 'The reduction of thought to a mathematical apparatus conceals the sanction of the world as its own yardstick. What appears to be the triumph of ... rationality, the subjection of all reality to logical formalism, is paid for by the obedient subjection of reason to what is directly given. What is abandoned is the whole claim and approach of knowledge: to comprehend the given as such ... Factuality wins the day. (Horkheimer and Adorno 1972: 26–7).

In spite of the 'dialectical' claims of Karl Marx, who pretended to have stripped eighteenth-century rationalism of its mechanistic features, his notion of reason is very much rooted in the Enlightenment tradition, to the extent that he believed that the historical process of productive forces is rational in itself and, therefore, emancipatory. This belief, the Frankfurt school asserts, is a fallacy, and Habermas, in particular, systematically addresses himself to this question.

The 'liquidation' of reason 'as an agency of ethical, moral, and religious insight' (Horkheimer 1947: 18) would not have been consummated during the course of the last centuries without a concomitant denaturation of the philosophical language and the language used in ordinary human affairs. Divorcing words and concepts from their respective intrinsic content, the Enlightenment triggered a process of corruption of speech[4] which has led to cultural decay. Horkheimer writes:

Language has been reduced to another tool in the gigantic apparatus of production in modern society. Every sentence that is not equivalent to an operation in that apparatus appears to the layman just as meaningless as it is held to be by contemporary semanticists who imply that the purely symbolic and operational, that is, the purely senseless sentence, makes sense ... Insofar as words are not used obviously to calculate technically relevant probabilities or for other practical purposes, among which relaxation is included, they are in danger of being suspect ... for truth is no end in itself. (Horkheimer 1947: 22)

Horkheimer sees the process of denaturation of language as a result of the deep socialization of the individual into the modern industrial system. In some pages of *Eclipse of Reason* he anticipates the thrust of what Riesman and his associates had to say in *The Lonely Crowd*.[5] Horkheimer describes the modern individual as a 'shrunken ego, captive of an evanescent present, forgetting the use of intellectual functions by which he was once able to transcend his actual position in reality' (Horkheimer 1947: 22). The modern individual has lost the capability to use language to convey meanings. He is rather able to express purposes. Accordingly Horkheimer refuses to accept the common 'behavior of people' (Horkheimer 1947: 47) in modern society as a basis for deciding the meaning of rationality. Without dissimulating his moral indignation toward modernization, he finishes his book *Eclipse of Reason* with the following statement: 'The denunciation of what is currently called reason is the greatest service reason can render' (Horkheimer 1947: 187).

The notion of rationality is also paramount in Habermas's works. He is mainly concerned with building a critical social theory as a tool for establishing the primacy of rational conduct in social life. Contrary to Weber, Habermas does not suspend ethical standards when addressing himself to the subject of rationality in modern societies. In the context of this chapter, it seems that Habermas's work becomes relevant to the extent that it deals with the following themes: 1) the restoration of the concept of an interest of reason, which although implied in Greek political thought, was made a central theme in the philosophical systems of the

German idealists; 2) a re-examination of Marx's views of history and especially his assumption that a rational society would necessarily result from the development of productive forces; 3) an inquiry into the political and psychological consequences of the grip of instrumental rationality on modern societies; and 4) the patternization of communication as a topic central to an integrative critical social theory. He leans toward a sort of integrative criticism.

Habermas delves into the mainstream of German idealism in order to examine rationality from a 'critical' viewpoint. He underlines that in Kant's transcendental philosophy 'the concept of an interest of reason already appears' (Habermas, 1971: 198). Pure reason in Kant's works has a practical interest in becoming incarnated in social life. Reason was conceived by Kant as endowed with causality. From its nature one can induce the notion of a good to be pursued in the realm of the individual as well as of social life. 'Reason prescribes an ought exclusively to rational beings,'[6] Kant said, and this thought was the whole theme of his *Critique of Practical Reason*, which, Habermas believes, contains the rudiments of a critical social theory. Moreover Kant is the root of German sociological thought in one form or another. Habermas relies on the Kantian heritage in order to develop a social theory aligned with the forgotten meaning of rationality. In one of his summations of Kant's thought he says that 'in reason there is an inherent drive to realize reason' (Habermas 1971: 201). In other words, reason has a practical interest which ought to be made effective in a society of rational beings. The problem is how to make pure reason practical in the social world. The responses to this question have varied. Hegel and Marx believed that pure reason would become congruent with the practical reason of everyday life in an Age of Reason which they assumed would be the necessary outcome of historical evolution. Habermas fundamentally questions this assumption.

After emphasizing the notion of an interest in reason in Kant, Habermas remarks, however, that Fichte gave to the question a treatment that is particularly pertinent to a critical social theory. Fichte developed the 'concept of emancipatory interest inherent in acting reason' (Habermas 1971: 198), which inspires Habermas to elaborate a typology of cognitive interests as criteria to differentiate several research orientations in the domain of science. Fichte's theory is particularly significant because it identifies rationality as the essential attribute of the enlightened human consciousness, i.e., a consciousness liberated from the dogmatism which ordinarily plagues all known forms of everyday life.

The notion of cognitive interest becomes a central tool to distinguish among several types of science. Habermas differentiates the sciences according to their research-guiding interests, to wit: (1) 'sciences (natural sciences) whose cognitive interest is technical control over objectified processes'; (2) sciences whose cognitive interest is a 'preservation and expansion of the intersubjectivity of possible action-oriented mutual understanding'; and (3) sciences subordinated to the emancipatory cognitive interest, i.e., which are to be considered as a tool to foster man's capability for self-reflection and ethical autonomy (Habermas 1971: 309–10).

The research-guiding interest of a critical theory of society is the emancipation of man through the development of his capabilities of self-reflection. However, in the model of established social science, technical control of reality is the basic research-guiding interest. This is to say that established social science has become scientistic by assimilating the method of natural science (more on this in the next chapter). Moreover it has tranformed itself into a means for legitimizing the institutionalized control of the natural world and human behavior. Effectiveness in controlling reality becomes the common criterion of validity in both natural and social science. Habermas strives for a social science on a different basis. He points out that the 'science of man ... extends in methodical form the reflective knowledge' and 'puts forth the claims to be a self-reflection of the knowing subject' and 'of the history of the species' itself (Habermas 1971: 613).

Habermas sees himself as continuing the Marxist theory. He claims that a critical social theory has to absorb Marx's contribution and to disentangle itself from his mistakes. Indeed Marxist theory is geared toward the emancipatory interest, which Habermas's notion of critical social theory shares. However, a fundamental correction has to be introduced in Marx's view of rationality and freedom. Marx supposed that freedom and rationality would be inevitable outcomes of the development of productive forces. Habermas remarks that this assumption has not been historically validated. He says: 'the growth of productive forces is not the same as the intention of the good life' (Habermas 1970: 119). The fact is that in industrial societies the logic of instrumental rationality which enhances the control of nature, i.e., the development of productive forces, has become the logic of human life in general. Even the private subjectivity of the individual fell captive to instrumental rationality. Capitalist development imposes limits to free and genuine communication among individuals.

Marx's assumption has been proved untenable by the simple fact that in the 'large-scale industrial society, research, science, technology, and industrial utilization were fused into a system' (Habermas 1970, 104) thus leading to a repressive form of institutional framework in which the norms of mutual understanding of individuals are absorbed into a 'behavioral system of purposive-rational action' (Habermas 1970: 106). In other words, in such an environment the difference between substantive and pragmatic rationality becomes irrelevant and even disappears. In fact the techno-industrial society legitimizes itself through the objective concealment of such a difference.

Identical in its position toward Marx is Habermas's reformulative approach to Max Weber. He explains Weber's concept of rationalization as follows: 'The superiority of the capitalist mode of production to its predecessors has these two roots: the establishment of an economic mechanism that renders permanent the expansion of subsystems of purposive-rational action, and the creation of an economic legitimation by means of which the political system can be adapted to the new requisites of rationality brought about by these developing subsystems. It is this process of adaptation that Weber comprehends as "rationalization"' (Habermas 1970: 97–8).

Habermas, however, deems it necessary to develop further the analysis of rationality, since in its present stage the industrial society is very different from the one Weber knew. Weber could address himself to the theme as a functionalist; however, today the question entails striking ethical connotations, which the theoretical effort of Habermas considerably highlights.

In a commentary on Marcuse, Habermas points out that at the present stage 'what Weber called "rationalization" realizes not rationality as such but rather, in the name of rationality, a specific form of unacknowledged political association' (Habermas 1970: 82). Moreover it seems an 'apologetical standard' (Habermas 1970: 83) in which norms of interpersonal relations in the private sphere and rules of systematically purposive-rational action become identical or lose differentiation, and therefore lead to a situation of systematically distorted communication among human beings.

The phenomenon of distorted communication has become a central concern of Habermas. He proposes a distinction between purposive-rational or instrumental action and communicative action or symbolic interaction. The first, subordinated to technical rules, can be proven correct or incorrect. The second i.e., symbolic interaction or communicative action, defines interpersonal relations as free of external compulsion, their norms being validated 'only in the intersubjectivity of mutual understanding of intentions' (Habermas 1970: 92). A fundamental thesis of Habermas is that in modern industrial society the old bases of symbolic interaction have been undermined by behavioral systems of purposive-rational action. In such societies symbolic interaction is possible only in very residual or marginal enclaves. What keeps a society working as a meaningful cohesive order is the acceptance by its members of the symbols by which it interprets itself. Symbolic interaction is the essence of meaningful social life, and thus, to use a word of Kenneth Burke,[7] symbolicity is an essential attribute of human action. Meaning in human and social life is obtained through the practice of symbolic interaction. But in the industrial society meaning has been subdued under the imperative of technical control of nature and accumulation of capital.

A consequence of the grip of instrumental rationality on modern societies is that systematically distorted communication prevails among people. Such communication becomes 'normal,' otherwise the repressive character of social relations would become evident. Habermas underlines the effect of political and economic factors upon communication patterns. The study of the repressive character of patterns of communication prevailing in modern societies requires a theory of communicative competence. One can postulate that 'every speech, even that of intentional deception, is oriented toward the idea of truth.' If so, what are the patterns congenial to such a language? Such speculation leads to the notion of the 'ideal speech situation' and the idea of the competent speaker. In fact 'communicative competence means the mastery of an ideal speech situation.'[8] Where intersubjective relationships are reified because of social constraints impinging upon them, communicative competence is very hard to achieve. In such an environment the competent speaker is exposed to misunderstanding, not to mention also that he may be considered eccentric. Habermas points out that 'on the strength of communicative competence we can by no means really produce the ideal speech situation independent of the empirical structures of the social system to which we belong; we can only anticipate this situation' (Habermas 1970a: 144). The ideal speech situation cannot materialize unless within its adequate social context.

ERIC VOEGELIN'S RESTORATIVE WORK

From the vantage point of contemporary standards of political and social science, Eric Voegelin's work appears as heterodox, obscure, and even upsetting. In his estimation political science as envisioned by Plato and Aristotle has never lost validity. He speaks as an expert in hermeneutics, not as a chronicler of ideas. From the hermeneutic viewpoint, what essentially matters in the classical texts is the experiences they articulate. Unless the reader engages in an effort to re-enact in his psyche those classical experiences, he cannot apprehend their meaning. The oblivion of the content of those texts which bears upon human life is more than an instance of misinformation; it is a symptom of the deformation of the human psyche and constitutes what Voegelin calls derailment. Voegelin considers the last five centuries of Western history a period of derailment and deculturation of mankind to the extent that they exposed it to a process of 'systematic confusion of reason' (Voegelin 1961: 284).

One can correctly speak of reason as a reality independent of our speech. Any attempt to speak of reason as if it were only a conventional language construct reflects a deformed state of the psyche. Reason was discovered by the mystic Greek philosophers. But this historical episode is more than an incident interesting enough to be registered in the chronicle of ideas; it rather begins a period of the formation of the human soul. With such a discovery the human soul acceded to a level of self-understanding in which it breaks the confines of the compact view of reality articulated in myth. To be sure the event did not change the structure of the human soul: it rather represents a peak moment in which man's consciousness of his psyche gains in luminosity and differentiation.

Compared with Voegelin's interpretation of classical texts, Weber's and Mannheim's statements on reason convey tenuous inklings into its nature and therefore their works illustrate cases of an arrested assessment of modern society. Without an uncompromising allegiance to reason, as Plato and Aristotle explain it, Voegelin indicates that there is no possibility of a scientific political theory. His vision of the healing process of our present ills assumes the perennial validity of the classical paradigm of the good society and rejects any kind of value-free and 'relationalistic' social science.

A rational society, asserts Voegelin, can be nothing but what classically has been conceptualized as a 'good society.' In his 'restorative' endeavor Voegelin claims that the Platonic and Aristotelian notion of the good society is by no means a historical curiosity, but a model essentially correct for evaluating any existing society. This claim does not involve a rigid allegiance to Plato and Aristotle, as if in their works the model were to be held up as a dogma. The case is rather that their understanding of the matter is theoretically valid, although the operational problems of building a concretely good society are always bound within specific contexts. Plato himself was very sensitive to contexts and did not admit of a single paradigm of the good society. He admitted that one should be flexible enough to consider 'second, third, and forth best paradigms' (Voegelin 1963: 38) – each legitimate within limits of given circumstances.

The following elements can be attributed to the classical notion of good society, as it is being restored by Voegelin (1960).

First, a good society is one in which the 'life of reason' becomes the paramount 'creative force.' In fact, even though this view reverberates in the conceptions of the French philosophers of the eighteenth century as well as in Fichte, Hegel, and Marx, it occurs in degraded terms, i.e., all these thinkers agree that to qualify a society as rational is to assert that it is good, but from their standpoint history, not the human psyche, is the site of reason. It is precisely this misplacement of reason which leads these thinkers to dodge the issue of 'good society.'

In Plato and Aristotle one fundamental expression of the life of reason is the continuous tension inherent in man's existence as open to realms of reality not included in history. To assume that the final explanation of this tension can ever be found in a theoretical system, or that it can be eliminated with the advent in history of a 'good society,' essentially constitutes what Voegelin would consider the 'gnostic' and immanentist fallacy pervading the modern frame of mind. Rationality in the substantive sense can never be a definitive attribute of society. It is directly apprehended in the human consciousness, not by social mediation. It moves the individual towards a continuous, responsible, and arduous effort to subdue his passions and inferior inclinations. In Plato and Aristotle there is an insoluble dualism between reason and society, which itself constitutes the precondition of freedom. Any sociomorphic solution of this dualism would therefore entail a deformation of human existence.

Secondly, it results from the foregoing that a good society is hierarchical. Although men as rational beings are potentially equal, for circumstances we cannot explain they are not identically able to stand the tension implied in the life of reason. The ablest in standing this tension constitute a minority, and a society is good to the extent that this minority carries out leading political functions. Voegelin does not hesitate to speak out against what he considers the mistaken democratic feelings of the present age. Sheer egalitarianism is against the life of reason, whose intrinsic requirements alone give rise to the 'good society.' In the 'good society' social differentiation between people has to be acknowledged and legitimized as a consequence of the diversity in the objective ability in human beings to bear the life of reason. In other words, status, wealth, race, and sex should have no place as criteria for allocation of authority and power.

A third point to be made regarding the issue under examination is that according to the classical view the 'goodness' of a society is conditioned by empirical circumstances such as resources and the size of the population. It is already implicit in the classical tradition that the state of productive forces bears upon the possibility of political equality as even a Marxian would acknowledge. Furthermore the reflections of Plato and Aristotle as to how the size of the population can affect the quality of a polity proved to be pertinent. And facts of contemporary social life, as for instance those Robert Dahl points out in the book *After the Revolution?*, bring support to Aristotle's view that beyond a certain size a society becomes prone to irrationalities of different sorts.

Finally, a fourth observation is in order, one which evidences the realism of the classical notion of a good society. Such a society can never be implemented once and for all. Its corruption starts in the very moment of its inception, since it is subject to the 'cyclical law of decline and fall' (Voegelin 1963: 39). It is not conceived as a paradise on earth, the eschatological end of alienation and contradictions between man and the world. The hope for a final, perfect, harmonious social stage is vain. In the classical conception of the good society there is no promise of a 'realm of freedom' such as was visualized by Hegel and Marx. Such promise is the essential characteristic of what Voegelin calls gnostic doctrines.

Voegelin attributes great value to Thomas Reid's view that in common sense there already is a 'certain degree' of rationality. He states that 'common sense is a compact type of rationality,'[9] and therefore social transactions based on an undistorted perception of reality are possible. If reason is part of the structure of human existence, then understanding and conversation between men is possible on the basis of their common participation in reality. Yet true, rational debate is becoming a very unlikely possibility in modern societies. In such societies the psyche of the average individual has been assimilated into the model of a closed self entirely included within mundane confines. Today man's skills for rational debate have been damaged by the prevailing language patterns, together with his assimilation into the existing social framework in which instrumental rationality has become rationality in general. In modern societies rational debate is possible only in a very few restricted enclaves. Even in the so-called intellectual milieux one is generally unable to entertain rational conversation. Voegelin states that the decline of rational debate is a late phenomenon in Western history. He sees in the past a 'period in which the universe of rational discourse was still intact because the first reality of existence was yet unquestioned' (Voegelin 1967: 144).

It is significant that in Saint Thomas Aquinas's time 'rational debate with the opponent was still possible' (Voegelin 1967: 144). On this assumption Saint Thomas believed himself able to persuade pagans and especially Muslims of the validity of Christian truth with rational arguments alone. Thus in the *Summa Contra Gentiles* Saint Thomas says: 'against the Jews we are able to argue by means of the Old Testament, while against heretics we are able to argue by means of the New Testament. But the Mohammedans and Pagans accept neither the one nor the other. We must, therefore, have recourse to natural reason, to which all men are forced to give their assent.'[10]

Today we may have difficulty understanding Saint Thomas. Not only reason but also a few key words have undergone the obliteration of meaning highlighted in this analysis. Language itself has been apprehended under operational standards of efficiency, a fact which bears upon the entire realm of human existence. When viability and expedience replace truth as the overriding criterion of language, there is scant, if any, room for the persuasion of people through rational debate. Rationality vanishes in a world in which means and ends become the only reference for human actions.

SOME CRITICAL COMMENTS

All of these scholars seem to agree that in modern society rationality has become a sociomorphic category, i.e., it is interpreted as an attribute of the historical and social processes rather than a force active in the human psyche. They all acknowledge that the concept of rationality is determinative of the approach to matters of social design. Yet they all are less than sufficiently systematic in presenting their views on these matters. In their critique of modern reason they assume various postures: resignation (Max Weber), relationalism (Mannheim), moral indignation (Horkheimer), integrative criticism (Habermas), and restoration (Voegelin).

Since an appraisal of Max Weber and Karl Mannheim has previously been presented, a brief assessment of the Frankfurt School and Voegelin is now in order.

There is merit in both Horkheimer's and Habermas's works insofar as they strive to lay bare the basic fallacy of Marx's view of reason as an attribute of the historical process. They both would question the assumption that the unfolding of productive forces would necessarily in itself lead to the advent of a rational society. Horkheimer seems to indicate that the moment reason is misplaced from the human psyche, where it belongs, and made an attribute of society, the possibility of social science is lost. Habermas emphasizes that in advanced industrial societies, the productive forces themselves ultimately are political constraints which shape human life at large. To overcome this condition Habermas suggests that room should be opened for politics and rational debate with their function of guiding the social process. Those are positive features of both Horkheimer's and Habermas's analysis.

However, Horkheimer's work is not much more than an indictment against modern society, which, while illuminating, fails to indicate where and how to move in order to find alternatives for the present theoretical and social malaise. It seems that Habermas is concerned with such alternatives, but they are presented in cumbersome, eclectic, and rather sociomorphic terms. Indeed, Habermas's notion of 'knowledge interests' (which is less original than it appears when one carefully reads Plato and Aristotle, as well as Kant) may serve to identify the reductionist bent of scientism. However, that notion itself is not completely rid of scientism, to the extent that for Habermas philosophy is absorbed in the threefold types of science he proposes. What he calls 'intersubjectivity' and 'mutual understanding' bear upon spheres of reality which necessarily escape the reach of a merely scientific approach.

His 'critical theory', understood as an integration of what he believes are valid insights found in the works of Kant, Hegel, Fichte, Marx, and Freud, seems too eclectic and is still pervaded by sociomorphic fallacies. Apparently Habermas holds the assumption common to Fichte, Hegel, and Marx that human emancipation may happen as a collective social event. In order to create the possibility of such an event, he goes so far as to propose 'the organization of processes of enlightenment' (Habermas 1973: 32) and resuscitates the Marxian idea of an enlightened mass practice. There is thus a sociomorphic overtone in Habermas's project of 'a theory designed for enlightenment' (Habermas 1973: 37) that promises existential enlightenment as a collective quality of mass conduct, when enlightenment has always been possible only at the level of the individual psyche. It is not accidental that he sees Freud's doctrine as a subsidiary element of that 'theory designed for enlightenment.' This is an indication that Habermas supports a motivational type of psychology which rules out the role of reason in the human psyche. Such flaws in Habermas's theoretical endeavor are somewhat intriguing, if not perplexing, because he seems to have a good grasp of classical political theory (Habermas 1973: 41–81).

While all of the authors previously analyzed deserve to be considered as critics of modern reason, only Eric Voegelin contends that modern reason expresses a deformed experience of reality. Accordingly he considers it pointless merely to try to conciliate or integrate ideas and doctrines predicated upon modern reason. The case is that all such ideas and doctrines obscure the poles of human existential tension. They express an attempt and even a dream of resolving within history, society, or nature the tension (*metaxy*) constitutive of the human condition. As reason implies the awareness of this tension, reason, in the modern sense, is a misnomer.

Voegelin is, in a peculiar sense, a psycho-historian. He asserts that the classical texts are not relics to be appreciated from an evolutionary viewpoint. Rather the insights articulated in those texts are perennially valid. The human condition is not finally explained in these insights, but the experience of reality which engendered them is paradigmatic. No effort at understanding reality is permitted to rule out the critical requirements of such types of experience. It is not accidental that no definition of reason is ever presented in Voegelin's work. Significantly the work in which he most systematically deals with this subject is entitled 'Reason: The Classical Experience.'[11] Here he interprets texts, explores and expands insights, and identifies some features of the deformed experience of reality throughout history. His entire work is an attempt to assess ideas and events from the standpoint of the classical experience. Precisely in this sense, and not in Erik Erikson's sense, it is a study in psycho-history.

Because Voegelin seems to constrain himself to the role of interpreter of texts and analyst of ideas and events, the prescriptive content of his work is indeed very broad. Even his statement on the 'new science of politics' is void of immediate pragmatic concerns. It is built on the interpretation and expansion of classical insights and the critical assessment of modern Western intellectual and political history. It seems, though, hard to accept the notion that political science can be exclusively spelled out only in such broad terms. After all, Plato and Aristotle, the basic sources of Voegelin's pursuit, included in the field of political science features of everyday life which completely escape his attention. In their texts criteria of action for designing and improving existing polities are offered so that even today, on the basis of their teachings, it would be possible to infer how they would go about finding the institutional means to overcome the problems of contemporary polities. Although Voegelin's statement on political science leaves the reader very much enlightened about its classical underpinnings, it does not provide an inkling regarding operational implications for contemporary society.

Moreover his reading of modern science may be lacking in some qualifications. For instance, not all of today's scientific pursuits are 'scientistic,' i.e., unaware of the particular realm of being where natural science belongs. It seems that the type of science implied in the investigations of A.N. Whitehead, Werner Heisenberg, Arthur Eddington, Michael Polanyi, and others is very much attuned to reason in the classical sense, as will be explicated in chapter 2, and perhaps even illustrates a new degree of the differentiation of consciousness of reality taking place nowadays. In fairness it must be said that in occasional conversations and lectures, Voegelin has indicated his sense of this event.

Also to be emphasized is Voegelin's neglect in systematically defining the meaning of some basic terms of his vocabulary, such as derailment, *kinesis*, gnosticism, text, theory, orientation and motivation psychology, compactness, and differentiation. So far, terms like the above are insufficiently elaborated in Voegelin's writings.

Finally, Voegelins' view of political science needs more qualifications than he usually cares to offer, since as it is articulated so far it may sometimes appear tainted with an over restorative character. No return to any historical mode of human existence can be implied in the idea of a true creative restoration of the classical teachings. This restoration consists in making the classical thinkers, through the appropriation of their insights, active partners in the contemporary scholars' pursuit of knowledge. Restoration of the classical conceptual legacy in this case is meant only to overcome its oblivion. The classical thinkers are not to be considered infallible canonic authorities. After all one does not have much to learn from the Aristotle who justified slavery, only from the Aristotle consistent with the definition of the human being as *zoon politikon*.

CONCLUSION

In fairness it must be said that modern social science is not wrong, as long as one realizes the precarious character of its main assumptions, to wit, that the human being is nothing but a reckoning creature and the market the paradigm according to which his associated life should be organized. Indeed modern social science was construed for the purpose of liberating the market from the fetters which, throughout mankind's history up until the rise of the commercial and industrial revolution, kept it within definite confines. What now undermines the theoretical validity of modern social science is its lack of systematic understanding of the specific nature of its assignment. For more than two centuries the narrow theoretical scope of modern social science has been the cause of its impressive operational and practical success. However, today the expansion of the market has reached a point of diminishing returns in terms of human welfare. Modern social science should therefore be recognized for what it is: a creed rather than a true science.

Today the outcomes of modernization such as psychological insecurity, degradation of the quality of life, pollution, and waste of the planet's limited resources hardly disguise the deceptive character of contemporary societies. The self-definition of advanced Western industrial societies as the carriers of reason in history is being undermined daily, and is in fact so widely disparaged that one wonders if the legitimation of such societies solely on the ground of functional rationality will in the short run continue to find believers in this world. Such a climate of perplexity may turn out to be the harbinger of a theoretical breakthrough.

This critique of modern reason is not undertaken as an inconsequential academic exercise. Its purpose is to pave the way for the development of a new science of organizations. Reason is the root concept of any science of society and organizations. It prescribes a design according to which humans ought to order their personal and social life. Throughout the last three hundred years functional rationality has bolstered the effort of centric Western populations to dominate nature and to enhance their productive capacity. This is certainly a great accomplishment. But there are now indications that such a success may be on the verge of becoming a Pyrrhic victory. The awareness of this situation is opening new avenues of intellectual pursuits.

Current organization theory gives a general normative character to the design requirement implied in functional rationality. Taking for granted the unlimited intrusion of the market system upon human existence, current organization theory is therefore theoretically unable to offer guidelines to create social spaces in which individuals can engage in truly self-gratifying interpersonal relationships. Substantive rationality postulates that the proper site of reason is the human psyche. Accordingly the human psyche is to be considered the referent for ordering social life as well as for conceptualizing social science in general, of which organization scholarship is a particular domain. The role of substantive rationality in the structuring of human associated life is the subject of chapter 2.

2

Towards a Substantive Theory
of Human Associated Life

Although modern social science generally and organization theory in particular fails to distinguish sufficiently between functional and substantive rationality, they are nonetheless cardinal categories of two distinct conceptions of human associated life. It is the purpose of this chapter to distinguish analytically these two conceptions. Such an exploration is now imperative because theories of organization and social systems design exclusively predicated on the modern conception of reason lack true scientific validity. As in the critique of modern reason in chapter 1, it is also necessary to begin this analysis with Max Weber.

One may argue that when Max Weber felt that he should characterize 'modern reason' he was proceeding as a historian. Instead of assuming a substitutive posture toward classical reason as Hobbes did, Max Weber implicitly warned that in modern times a new meaning was being bestowed upon the word reason. He did not dismiss the previous meaning of reason as an anachronism. Indeed Weber as well as Hobbes wanted a type of social science entirely committed to a task peculiar to a given historical epoch. But by distinguishing between *Zweckrationalität* (formal rationality) and *Wertrationalität* (substantive rationality), Weber intimated that either one or the other could be the referent of theory-building. In fact he chose to develop a type of theory mainly predicated upon the notion of functional rationality. Although the biographical and historical background of Weber's choice would constitute an interesting and relevant matter of investigation, it is beyond the concern of this chapter. Nevertheless I submit that a substantive theory can be formulated on the basis of what Weber did not but likely would have said had he lived in the present historical circumstances.

It is Max Weber's contention that although social science is value-neutral, values embraced by a society are themselves criteria which indicate what issues are relevant to a particular form of human associated life during a certain historical period. He would then admit that when the value premises of a certain type of associated life become themselves the factors of a collective malaise, the social scientist cannot legitimately discard them as foreign to his discipline. Rather, from Weber's vantage point, the social scientist could only focus upon those values to show their empirical consequences. The social scientist as such should not utter value-judgments since values are subjective or demoniacally founded.

Weber's position, however, is not without contradiction. If values are simply demoniacal and do not have ojective grounds, then the analysis of the consequences of their adoption by individuals is nothing but a futile exercise in abstraction. Such an analysis would make sense only if it were undertaken on the hope that the individual could be persuaded to make an objective, rational value-judgment.[1] This contradiction in Weber's position is reflected in his work and life. He pretended to have studied, *sine ira ac studio*, the syndrome of formal rationality, but nevertheless expressed his regret about the outcome of such a syndrome – a world of 'specialists without spirit, sensualists without heart' (Weber 1958: 182).

Max Weber lived in a historical context in which formal or functional rationality largely replaced substantive rationality as the major criterion for ordering political and social affairs. He took for granted such a replacement, and refused to build social science on the notion of substantive rationality. Today, however, it is more difficult than in Hobbes's and Weber's times to dismiss the viability of a substantive theory of human association because it is now evident that value relativism has led associated life into an intellectual and spiritual dead end. Accordingly the question this chapter will address is whether substantive reason should be the cardinal category for thinking about political and social matters and, if so, what type of theory would correspond to such thinking. It will be the purpose of subsequent chapters to discuss resulting emergent social structures and policy implications.

There are three general qualifications which highlight the distinctions between substantive and formal theory of human associated life.

First, a theory of human associated life is substantive when reason in the substantive sense is its cardinal category of analysis. Such a theory is formal when reason in the functional sense is its cardinal category of analysis. To the extent that substantive reason is understood as a prescriptive category, substantive theory is a normative theory of a specific kind. To the extent that functional reason is merely a definition or logical construct, formal theory is a nominalist theory of a specific kind. Concepts of substantive theory are insights into the process of reality, while concepts of formal theory are merely conventional language tools descriptive of operational procedures. The question *What is rationality?*, which requires direct attention in the domain of substantive theory, has no role to play in the domain of formal theory; here the question is rather *What shall we call rationality?* The question would be answered in the latter case by a statement in which a combination of words[2] essentially constitutes the referent for analytic pursuits.

Second, a substantive theory of human associated life has an age-old existence and its systematic elements can be found in the works of thinkers of all times, past and present, attuned to the common sense meaning of reason, although none of them has ever employed the expression 'substantive reason.' Indeed it is because of peculiarities of the modern period whereby the concept of reason has been overtaken by functionalists of various persuasions that we now need to qualify the concept as 'substantive.' One basic finding which has resulted from the teaching legacy of classical thinkers is that it is rational debate, in the substantive sense, which constitutes the essence of the political way of life and is an essential requisite for the sustenance of any well regulated human associated life at large.

Incidentally, what the field of economics and more specifically the field of economic anthropology currently refers to as substantive theory[3] is only subsidiary to the present analysis. Today's debate between substantive and formal economic theorists refers to the nature of the economic phenomenon, the market, and its theoretical implications. Karl Polanyi, the seminal substantive economic theorist, points out that formal concepts drawn from the specific dynamics of the market are at best valid as general tools for societal systems analysis and design only in a capitalist society during a period in which the market is relatively free from political regulation. Formal economic theorists claim that those formal concepts are universally valid. Polanyi correctly states that as the economy has always been 'embedded in society',[4] the capitalist society has to be understood as an exceptional case and not as a paragon of social and economic history. The fact that this observation requires statement and elaboration is itself indicative of a peculiar historical condition. Premodern political theorists did not need to stress this point because they were never exposed to such a condition. However, they meant the same thing as Polanyi when they stipulated that man's gregarious life must be politically regulated.

The third and final qualification is that substantive theory as conceived here implies an ethical superordination of political theory upon any eventual discipline bearing upon human associated life.

In order to clarify analytically the distinction between substantive and formal theory of human associations an explication is required of the statements presented in the accompanying table. Discussion of these points comprises the rest of this chapter.

THEORY OF HUMAN ASSOCIATED LIFE

Formal	*Substantive*
1 Standards for ordering human associations are socially given.	1 Standards for ordering human associations are rational, i.e., self-evident to the individual common sense apart from any particular socialization process.
2 A fundamental condition of social order is that the economy becomes a self-regulated system.	2 A fundamental condition of social order is the political regulation of the economy.
3 Scientific study of human associations is value-free: there is a dichotomy between values and facts.	3 Scientific study of human associations is normative: the dichotomy between values and facts is a false one in practice and tends to produce distortive analysis in theory.
4 The meaning of history can be captured by knowledge which discloses itself through a series of determinate empirical-temporal stages.	4 History becomes meaningful to man through the paradigmatic mode of the polity's self-interpretation. Its meaning cannot be captured by serial categories of thinking.
5 Natural science provides the theoretical paradigm for correctly focusing upon all issues and questions posed by reality.	5 Proper scientific study of human associations is a type of inquiry in its own right, distinct from the science of natural phenomena.

1 / THE MODERN TRANSVALUATION OF THE SOCIAL

Modern social science could not have been constituted until the political element in peoples' lives was reformulated and subsumed under social interests.[5] For the purpose of illustrating this point one might consider the teaching of classical thinkers. In fact they never envisioned sociality as a condition peculiar to man, but one he shared with other animals.[6] Classical thinkers were concerned with that characteristic of man which makes him a *sui generis* creature not completely included in the natural realm, and they recognized that this characteristic is man's awareness of the activity of reason in his psyche. By exercising reason and living according to its ethical imperatives, man transcends the condition of a purely natural and socially determined being and becomes a political actor. Aristotle's definition of man as a 'political animal' (*zoon politikon*) is only comprehensible in this light.

Aristotle and the classical thinkers in general conceived of sociality as a herdlike quality unworthy of political man. In the political domain man is meant to act on his own, as a carrier of reason in the substantive sense. In the social domain, by contrast, the concern with 'mere life' prevails, and he acts as a 'reckoning' creature, i.e., as an economic agent. Reason in the sense of a 'reckoning' skill is also implied by Aristotle in the *Politics* and the *Nicomachean Ethics*. It is required for the management of the household (*oikos*) where the economic welfare of the group dictates what course of action shall or shall not be taken. This kind of social conduct is limited to its proper enclave. It does not belong to the political domain where the individual may exercise his concern for the enhancement of the good character of the whole, and not simply for survival.

Aristotle was of course aware that the paradigm of the best political way of life, open to the full command of substantive reason, could only obtain in history by chance. He knew that no polity is forever guarded against the undermining influence of social interests. But where these practical interests constitute the only criteria for human actions there is no political life at all.

It is fashionable for many to offer a self-interpretation of the development of social science as the consequence in the modern historical period of an increasingly sophisticated conception of rationality and its application to increasingly diverse phenomena. The historical bias of this assumption will be treated more fully in Section 4 of this chapter. For now it is sufficient to point out that the very idea of a social science, predicated on the assumption that the individual is fundamentally a social being and that his virtues are to be assessed by criteria socially given, was inconceivable to Aristotle and the classical theorists in general. Modern social science implies that society, in unfolding itself as a purely natural association, generates the standards of human existence at large. Such a transvaluation of the social, of which modern social science is an outcome, has occurred in the last three centuries of Western history. It is Hobbes who systematically paves the way for this transvaluation by attacking the common sense notion of reason and proposing alternatives to it. The moment the human being is reduced to a reckoning creature, it is impossible for him to distinguish between vice and virtue. Society, then, becomes his only preceptor and, not surprisingly, pain is equated with evil and pleasure with the good.

One of the most striking documentations of the bewilderment of Western man in the peak moment of the transvaluation of the social is Bernard Mandeville's *The Fable of the Bees* (1714), first published as *The Grumbling Hive* (1705). In 1723 he published 'A Search into the Nature of Society,' an essay which is very much a theoretical justification of his fable. Mandeville compared society – in particular the British society of his time – with a beehive. One could summarize the basic argument of his work as follows: if the social becomes the all-inclusive criterion for ordering human existence, then vices, pride, selfishness, corruption, fraud, greed, hypocrisy, and injustice turn out to be virtues. Although he indicates that virtue is beyond human capability, Mandeville may be considered a moralist *malgré lui* if one interprets his fable as portraying the consequences for human existence at large of exempting society from political regulation. The ambiguity of Mandeville's thought largely explains why, through subtle and contradictory ways, he influenced eminent writers and philosophers of his time, including Adam Smith, who, however, disavowed Mandeville's influence.[7]

It is no accident that the idea of a 'social science' obsessed Scottish philosophers in England during the eighteenth century. The notion of 'social science' pervades the speculations about human nature and the institutional phenomenon undertaken by Adam Ferguson, David Hume, Francis Hutcheson, Lord Kames (Henry Home), Lord Monboddo (James Burnet), Adam Smith, and Dugald Stewart.[8] However different these men were in ideas and talent, they nevertheless had allegiance to a certain community of assumptions. One of these assumptions was that social science could be possible as the study of society's reason. For most of them reason is a characteristic of society rather than of the individual. They consummated the sublimation of reason in the sense that it no longer was to be conceived as mediated by the individual, but as something cognate with society and nature. They all conceived of rational laws governing society and nature, in spite of the fact that they agree that passions rather than reason move the human being to act. One usually attributes the idea of the 'invisible hand' to Adam Smith, but actually it was suggested in the writings of these men. The work of such a 'hand' manifests itself in society and nature and it is the role of science to discover and articulate how this transpires. For instance, Hutcheson, a friend and professor of Adam Smith, wrote in 1728 of a 'superior hand' as a providential force which affects both human beings and beasts through their 'instincts.'[9]

To Mandeville's credit there is no ambiguity in David Hume and Adam Smith when they justify the all-inclusiveness of society. For Hume and Smith, sociality replaces reason in prescribing how man must live. Hume considered the human being as a creature completely included in society. The 'merit' of his actions 'do[es] not derive ... from a conformity to reason, their blame from a contrariety to it' (Hume 1973: 458). It is his feeling of belongingness to society that constitutes the grounds for his moral conduct. Sympathy, the individual's sensitivity to others' approbation and blame, plays a fundamental role in the development of his moral sense. Standards for ordering human life are themselves social, and in the last analysis consist in the 'interests of society,' which Hume claims to be the main referents for distinguishing between virtue and vice (Hume 1973: 579). For Hume, order in human associated life is an outcome of anonymous processes involving forces and activities independent of the individuals' rational deliberations in the substantive sense. Rather what is social is necessarily moral. As with other Scottish philosophers, Hume intends to elaborate the philosophical groundwork for formal Western social science.

Today's prevailing currents of thought in formal social science, whether in its standard 'established' terms, or under its Marxian and Neo-Marxian guises, rely on a sociomorphic view of man. This view reduces the human being to nothing more than a social being. Hence the individual's full actualization is understood as his total socialization, either under the conditions now given, or in a future enlightened social stage. For instance, economic motivation is considered the paramount feature of human nature, and formal economic theory claims that the market is the cardinal category for comparing, assessing, and designing social systems. Thus it is no accident that 'established' social scientists recommend to 'third world' countries the mass practice of a certain type of organized enlightenment, intended to teach 'achievement motivation' or its equivalent to the people living in that area. This conceptual outlook implies that countries of the 'third world' can solve their problems only if they become market-centered societies. The Marxian and Neo-Marxian theorists realize the *naiveté* of this particular point of view. They underscore instead the imperative of changing human nature according to the model of socialist society, which, incidentally, they never conceptualize except in vague terms. Accordingly essential pieces of their universal blueprint are socialization of the means of production and of economic surplus, socialist planning of the structure of production and consumption, and egalitarian Marxist education as organized mass enlightenment.

Since, for both currents of thought, human nature lacks standards appropriate to itself, the measuring stick for assessing and designing social systems ultimately is social itself. Therefore formal social science can never be a critical theory, as some writers claim today, unless the 'critical theorist' secretively believes that only he is able to withdraw from the socialization process and enunciate verdicts on the current stage of human nature. The critical theorist's personal disappointment with the prevailing state of things may induce him to indulge in an oracular and demiurgic type of action. Ultimately there are no significant points of contention between 'established' Western social scientists and Marxian and Neo-Marxian 'critical theorists' because they both exemplify the modern transvaluation of the social and accordingly agree that a theoretical predicament is inherently embodied in a societal praxis: today's praxis of advanced industrial society or tomorrow's praxis of the enlightened masses. Hence it is vain and self-defeating to attempt to overcome the ideological fallacies of Western social science and at the same time preserve its formal character.

2 / POLITY AND SOCIETY

The full-blown paradigm of the Scottish idea of social science first appeared as 'political economy' in the works of Adam Smith. Proposed as a truly general social science, 'political economy' conceives of order in human associated life as a result of the free interplay of its members' interests. Thus the delimited enclave which Aristotle described as the order of the household becomes explicitly equated with human associ-

ated life at large by the British founders of social science. Political economy and formal social science conceptually legitimize the exemption of
the household economy from political regulation.[10] Hence commerce
becomes the essence of society, and human nature is defined in conformity with the qualifications of man as a merchant.

Paradoxically, however, there is a commonality between substantive
theorists and Western formal theorists: they both imply that reducing the
individual to a purely social being is equivalent to claiming that his main
concern in society is his sheer self-preservation. Both would claim that
society is a specific type of human collectivity in which immediate,
practical interests constitute the individual's basic referents for relating
with others. Formal theorists attribute a dynamic of its own to such a
collectivity. This dynamic guides the interplay of the individuals' interests toward an order of the whole of which none individually is
conscious. Accordingly, for the formal theorists politics is the articulation
and aggregation of interests and should either minimally intervene in the
free dynamics of society (liberalism), or be the expression of society acting
as a whole according to its laws (socialism). Thus one might say that the
expression 'political economy' contains a contradiction in terms.[11] It
misplaces an attribute (the political),[12] which only the human being can
exercise, by considering the economy as a corporate political actor itself,
i.e., as a whole animated by a purpose of its own, one that ultimately
coincides with the good of all men. Thus the idea of social science implies
a reduction from below of human associated life, since sociality per se is a
defective dimension of human existence.

One might inquire into the historical circumstances which constitute
the background for such a transformation. Since the sixteenth century a
number of events, such as the discovery of new parts of the globe and the
expansion of the area of transportation by sea – thanks to the initiative of
European navigators – as well as the multiplication of commercial and
industrial ventures, precipitated the emergence of a new attitude toward
material prosperity in Europe. By trading with other parts of the world,
some European countries found a way of increasing their wealth at an
unparalleled rate. In the last analysis what is called the commercial and
industrial revolution was a series of events implying that material
prosperity and wealth could be created or fostered by systematic human
deliberation. In the premodern centuries material prosperity and wealth
were outcomes of human deeds, but these deeds were intelligent transactions with nature as given. Man's needs were considered limited. The

productivity of nature was to be obtained by man's collaboration with nature's self-generated processes, and not by systematic escalation of these processes through technological means, irrespective of nature's thermodynamic constraints.

An exploitative posture towards nature would appear ethically vicious to the premodern and non-modern frame of mind. Scientific technology existed long before the so-called industrial revolution without necessarily dominating nature. In Greece, for instance, inventions attributed to Hero and Archimedes testify to an advanced stage of technology. But the application of technology to production was limited for political and ethical reasons. Technology, for the Greeks, should neither be the concern of a free man, nor violate nature's self-generated processes.[13] It is as if the Greeks and other ancient people were aware that the economy was a subset of the biophysical system.

Consumption within the limits of finite human needs and limited production constituted the goal of an instituted economy in premodern societies. In the thirteenth century Saint Thomas Aquinas reiterates this view. Following Aristotle he cautions against the proliferation of needs socially induced, and develops a distinction between 'natural' and 'artificial' wealth. The former 'serves to remove from man his natural deficiencies such as food and drink, clothing, vehicles, shelter, and such like.' 'Artificial wealth,' for instance, money, has been invented by 'human art ... in order to serve as measure of things exchangeable.'[14] Late in the seventeenth century this orientation still echoes in Pufendorf's writings.[15] He shares with the premodern thinkers the point of view that value in use should determine the activities of the economy. But exchange value, not value in use, constitutes the goal of a modern economy. Thus by legitimizing the pursuit of wealth for its own sake, this inversion necessarily implies the emancipation of the economy from political and ethical regulation.

Substantive political theorists realized that a good polity could not exist below a certain level of the means of subsistence. They considered the creation of such means as a necessary condition, rather than the goal, of the polity. However, since the moment that material prosperity was assumed to be possible for all (i.e., an assumption characteristic of the so-called industrial revolution), wealth became the cardinal goal of the polity. Throughout the centuries of commercial and industrial development, political theory is accordingly redefined. Its main concern became material prosperity rather than the goodness of human association. The ethical standard inherent in substantive political theory was replaced by the moral justification of the individual's immediate interest. Contemporary social science is the culmination of this process.

There was a moment in which the novelty of 'society' in Europe was felt and vividly discussed. For the next few paragraphs let us write 'society' between quotation marks, in order to underline its peculiarly modern meaning. Hegel, an admirer and careful reader of Adam Smith, focused upon 'society' as a historical event.[16] He salutes the event as a progression of freedom, since in his view 'society' is the arena where the universal is meant to finally consummate itself.

Although influenced by Hegel, Lorenz von Stein looks at 'society' rather as a historian. His book, published in 1850, *Geschichte der Sozialen Bewegung in Frankreich von 1789 bis auf unsere Tage* (translated into English under the title *The History of the Social Movement in France, 1789–1850*), significantly starts with a chapter on 'The Concept of Society and Its Dynamic Laws.' In it he highlights that in his generation 'society' denotes an order of 'phenomena which had formerly remained unrecorded in everyday life as well as in science' (von Stein 1964: 43). Lorenz von Stein undertakes his historical analysis in order to legitimize the 'new science,' i.e., 'social science.' In substantiating his effort he expresses his awareness of what the 'new science' fundamentally implies. Thus he writes:

There is something within the state operating against it. This something is *society* [italics in the original]. Does society adhere to a different principle of existence than the state? If so, what is the principle?

For centuries, many great men have tried to formulate the principle of the state, but nobody thought of the possibility that there might also exist a principle of society. And, yet it does ... Interest, which is the center ... of all social motion, is the *principle of society* [itals in the original] (von Stein 1964: 45, 55).

As with Adam Smith in the case of the economy, Lorenz von Stein welcomes the exemption of 'society' from political regulation. But this position was not shared by all German thinkers of his time. One of them, Heinrich von Treitschke, whose participation in German public affairs may deservedly be criticized, nevertheless may be called upon to illustrate the tension between the 'new science' and political theory. In Treitschke's view human communities can only subsist when provided with some form of state. In other words, since 'society' is unable to spontaneously give form to itself, such an ordering assignment belongs to the state. Thus Treitschke regretted that 'everything which our century terms Liberalism tends toward the social view of the state' (von Treitschke 1963: 29). From his viewpoint the culmination of such trends would lead to the disintegration and collapse of associated life. What he calls the 'social view,' the 'exclusively social attitudes of mind,' the 'purely social outlook' (29–30) implies a reference to the thesis of a 'social science' independent of political science. Assuming a somewhat polemical position regarding this thesis, he asserts that 'when our century claims that the study of social conditions is a new thing ... it exhibits a strange self-conceit' (32). Treischke explicitly denies the possibility of 'social science,' since political theory (in the substantive sense) has always properly dealt with the order of phenomena which the 'new discipline' pretends to define as its exclusive domain.

Today one hardly realizes the fundamental question implied by the rise of the 'social' and 'social science.' In its liberal and socialist models, formal social science conceives of human associated life as ordered by interest, which is the same as to assume that the principle of 'society' is the essential normative standard of human existence at large. In other words, by diffusing the political element in human associated life, formal social science is oblivious to any kind of substantive regulation bearing upon the economic process.

3 / THE DICHOTOMY BETWEEN VALUES AND FACTS

In focusing upon the dichotomy between values and facts, I here investigate the historical circumstances which gave rise to it rather than its philosophical underpinnings.

When the individual is defined as a purely social being, order in his life is supposed to be extraneously granted to him. The world from which this order emerges is an arena in which he endeavours to maximize his gains. The order of society is possible to the extent that its members, on the basis of a calculus of consequences, regulate and limit their passions so as not to jeopardize their immediate practical interests. Society is the market writ large. Human values become 'economic values' in the modern sense, and all ends are of the same rank. In the last analysis, as we shall see, the dichotomy between values and facts is only valid when the total inclusion of man into society is taken for granted.

Moreover in a market society the concern with matter as the stuff of which things are made prevails upon or even rules out a concern with their nature or intrinsic ends. The market is blind to the intrinsic ends of things and considers things and the individuals themselves, converted into the labor force, as 'facts,' to wit, as factors of production. In consequence, contemporary disciplines such as economics, which take for granted the market-centered society, have to be value-free, and exclusively concerned with 'fact.' Implied in these disciplines is the claim that values are simply aspects of human subjectivity. They are, at best, to be considered as exogenous or secondary qualities of things, not as their properties. Thus they cannot be the object of cognitive assessment. From the analytic viewpoint cognitive and normative statements then become mutually exclusive. It is of interest to note that such a dichotomization is reflected in the research interests within even the leading departments of social science in universities in this country today.

The predictive power of formal value-free social science must be acknowledged indeed, but we must realize that it is predictive only as long as the circular causation linking the market and the behavioral conformity of the individual to it remains undisturbed. However, the moment that such a conformity, for reasons beyond the scope of this analysis, is open to question, as is true in our time, a psychological resistance emerges directed against the unregulated dynamics of a market-centered society. Such opposition undermines the predictive power of formal value-free social science because the person tends to become more than a totally socialized being. This resistance triggers a normative trend of thought of which a substantive theory of human associated life is intended as an initial articulation.

Left to its own dynamics, the market system works against the constitution of associated life understood as a community of men and women. This fact has even been acknowledged by economists themselves. For instance in 1913 Philip H. Wicksteed depicted the 'industrial world' as an 'organization for transmuting what every man has into what he desires, wholly irrespective of what his desires may be' (Wicksteed 1913: 258). To counteract this effect he envisioned a type of economics subordinated to 'ethical considerations' (260), because 'the sanity of men's desires matters more than the abundance of their means of accomplishing them' (260). It is also his contention that 'the market does not tell us in any fruitful sense what are the national, social, or collective wants or means of satisfaction of a community for it can only give us sums' (260). In consequence he states that 'the economic laws must not be sought and cannot be found on the properly economic field' (260). What is regrettable is that such an insightful understanding of the nature of the market so far has not constituted itself into a coherent conceptual corpus convincing enough to replace the formal type of thinking still dominant in the conventional milieu of the economic profession in particular and of social science in general.

4 / SOCIAL SCIENCE AS A SERIALIST IDEOLOGY

The assumption that history discloses its meaning through a series of empirico-temporal stages is common to standard liberal academic as well as Marxian and Neo-Marxian theorists. Implied in this common asumption is a view of time peculiar to the Englightenment which continues to predominate in contemporary Western modes of thinking. In the writings of the epigoni of the Enlightenment the time in which human nature supposedly actualizes itself consists in seriality. Through distinct qualitative degrees of actualization, corresponding to different steps occuring in an ascendent serial kind of time, human nature changes its structure. Moreover in this Enlightenment perspective there is a culminating historical moment in which human nature reaches its final perfect stage.

To be sure the serial view of human existence in history has comparative diachronic and synchronic implications. When assessed against the structure it is supposed to reach in the culminating stage, human existence in past historical periods is of course deemed defective. And insofar as not all contemporary societies have arrived simultaneously at the same degree of advancement, human existence in those less developed societies, lagging behind the more advanced or even historically terminal ones, is also necessarily defective. For instance the notion of the 'third world' reflects the serial view of present history, since it assumes the 'second' and the 'first.'

Under the influence of the Enlightenment's serial mood several authors pretended to have understood the patterns according to which history unfolds itself. For instance, Condorcet, Turgot, and Saint-Simon felt summoned to lay bare the necessary stages of history. Contemporary social science is an offspring of the intellectual climate of which these authors are representative. In the *Wealth of Nations*, for example, where the conceptual corpus of political economy is presented, Adam Smith reinterprets history, describing the commercial society as its culminating stage. August Comte and Karl Marx undertake similar endeavors. They see in their own epoch the imminent culmination of history, defined, respectively, as the positive age and the socialist society.

Today the serialist mood continues to be characteristic of social science and makes itself evident in the way social scientists focus upon themes such as 'social change,' 'social stages,' 'modernization,' 'development,' 'post-industrialism,' 'advanced industrial society,' and 'socialism.'

These terms have been undermined by contemporary events such as the disenchantment with industrialism, the social malaise characteristic of 'advanced' societies, and the depletion of limited resources and pollution of the environment. Consequently they have become the object of a growing revisionistic literature. For instance, reacting to these circumstances, some theorists have undertaken a tinkering endeavor to rescue the notions of 'modernization' and 'development' from their historical embeddedness. In fact the notion of 'modernization' generates so many perplexing questions that it is about to be discarded from the language of formal social scientists. A number of writers are now trying to reconceptualize development not as meaning the unqualified increase of GNP, but essentially as denoting the qualitative enhancement of the human environment, and mainly as social and economic equalization. Impressed with the fact that successive programs of modernization and development implemented in the 'third world' have not changed its dependent status within the present frame of the international division of labor, some Neo-Marxists argue *post hoc* that those programs have always been concocted to serve imperialist intents.

Significant as these views certainly are as indications of the breakdown of modernity and developmentalism, still they miss the point. That is, their representatives have not sharply confronted the serial adventist mood and the pseudo-theoretical underpinnings of the standard formal type of social theory; they still remain caught within its meta-ideology, concealed as structural-functionalism, Hegelian dialectics, Marxism and Neo-Marxism, and the different mixes of these trends with phenomenology and/or existentialism.

From the standpoint of the comparative criteria of these trends of thought, such as, for instance, pattern-variables and developmental stages, the different societies of the contemporary world are classified along an Indian line pointing toward the so-called advanced or Enlightened society. In fact these criteria are epistemological traps and disguised ideologies that foster the misunderstanding of these societies and misdirect them from their critical imperative of self-reconstruction. Policies derived from these criteria work in practice to escalate the Westernization of the entire world. One of the results of this process in which the so-called 'third-world' nations are caught is the degradation of their internal structures. The feeling of relative deprivation which plagues especially the middle sectors of these nations is one of the primal factors hindering their self-reconstruction. The resolution of the problem that this feeling generates is conceptualized via the serialized category of developmentalism, in its standard or Marxist interpretations. This adventitious mood, rather than the scarcity of resources, constitutes the fundamental obstacle to their cultural, political, and economic self-articulation. In order to overcome this entrapment a break with the social ideology of the West is imperative.

One can conclude that such a break is imperative if present so-called underdeveloped nations are to find the way out of the process in which they are driven. But the point I wish to stress here is that the terms of this break cannot be found through any refurbishment of the serialist Western ideology. Such a break is not likely to occur unless the peoples' will be activated to build immediately out of what they already have a rational society, understood substantively and without current serialist and futurist connotations. The beginning and the end of history are not serial categories. Rather their meaning is apprehended through compact experiences of time. Before the modern Western period they have been experienced as immediately present to any society through its self-understanding as a microcosmos and have ordered people's lives here and now. In the past as well as now, however, only under hegemonic domination do societies eventually succumb to the serialist mood of relative deprivation. This episodical condition of human nature is not to be understood as human nature itself. (Further consideration of the concept of time and implications for social systems will be presented in chapter 8.)

Breaks have been undertaken by troubled and captive peoples in the past, as for instance when Abraham left Ur and Moses left Egypt.[17] In these cases the breakthrough was preceded by (or simultaneous with) a horizontal movement in the historical space. But in the planetary institutional frame of our time there are no more horizons for an exodus in the horizontal sense. If a historical break is to happen in our time it will have to assume the unprecedented character of a pure exodus in compact vertical time, i.e., through the change of peoples' hearts, their orientation toward reality, and criteria of perception and definition of their needs and wants. The rhetorical image of curtains associated with the Soviet and Chinese experiments may be indicative that in the present people can still be mobilized to attempt a breakthrough, although these 'communistic' experiments can certainly be considered as abortive, since they do not go beyond the serialist mood of Western ideology. Whether a genuine exodus from the West is feasible today is a matter for a type of discussion which exceeds the scope of this analysis.

5 / ON SCIENTISTIC SOCIAL SCIENCE

Formal social theory is scientistic. That is, it assumes that the correct account of reality can only be articulated according to the model of the technical language of natural science. In this view reality is narrowed down only to what can be operationally verified. Scientistic social science itself is an offspring of a serial posture toward reality.

A sound argument against scientistic science does not underestimate the relevance of operational questions. It simply asserts that method and technique are not standards of truth and proper scientific knowledge. The findings of scientistic science may eventually be corroborated, although it should be pointed out that they are restricted to a level of reality whose boundaries must be acknowledged. To consider this mode of knowing as the paradigm of knowledge in all realms of reality is precisely what Whitehead called the 'fallacy of misplaced concreteness.'

To be sure natural scientists themselves do not necessarily share the assumption of scientistic science. Einstein, for instance, seems to avoid indulging in sheer scientism. It is no accident that he claimed that 'it is ... theory [not method, A.G.R.] which decides what can be observed.'[18] This statement is significantly quoted by W. Heisenberg in an essay in which he tries to conciliate Copernicus, Galileo, Newton, and physical science in general with the classical tradition. He sees the historical trajectory of physical science not as radically discontinuous with the classical tradition, but as a 'history of concepts' (Heisenberg 1974: 56), an increasing differentiation of the knowledge of stable structures of reality. He equates mathematical constructs with Plato's archetypal forms (55). An even stronger Platonic accent is characteristic of Eddington's view of physical phenomena. For him the solidity which the 'untaught' eye sees in external nature is illusory or 'shadowy,' and science itself is nothing more than a symbolic type of knowledge of reality (Eddington 1974: xv-xvII and 318).

Formal social science, particularly in its behavioralist persuasion, is scientistic in the sense already elaborated. Its bias has been the object of several recent critical studies. They all indicate that the functionalization of its language and its method-centered orientation ultimately convert social science into a disguised form of ideology and technology. Illustration of this assessment will be provided here by a brief account of the behavioralist notion of political order and cross-national criteria of comparison.

Formal political science is 'apolitical' in the sense that it completely loses sight of the qualitative distinction between political life and social life. Indeed it equates social gregariousness with political order. The result of this confusion is abolition of the political element in human associated life. For instance it is David Easton's claim that society 'is the most inclusive social unit we know.'[19] He further develops this view as follows:

Political activity is vital in a society. But so are economic activity, social structure, and the like. A society has numerous aspects and it is unlikely that men could pursue their purposes without providing for goods and services, for example, as well as for the authoritative allocation of valued things. Marx seemed to argue for the primacy of economics, some social scientists have urged the dominance of culture, or of personality and motivation. The fact is that at the general level of societal existence, each major area of human activity contributes its share in a totally interactive process. Politics infuses ('formative principle'?) all of life; but so do economics, culture, motivation and the like. A multicausal, interactive interpretation of society seems more helpful in understanding the way it operates than one that insists on some kind of primacy, out of hand, for one or another social aspect. We may grant that empirically, at given times and places, one or another aspect of society – the political, economic, cultural, psychological or structural – may indeed attain a selective emphasis for special reasons. But it is difficult to imagine a kind of society in which each of the aspects mentioned could not be said to be 'formative' in some significant sense. (Easton 1973: 294)

Easton's statement is representative of the bewilderment prevailing among formal political theorists. By sharing his idea of the political element as belonging to the same level as the economic and the social, these theorists equate political order to control of gregarious life, regardless of the nature of its normative principles. For them political order exists in a society as long as it has institutional capabilities to induce the citizens to conform to its given standards, or to articulate and aggregate competing interests and thus persist in a stable state. In line with this purely operational view of political order an author feels safe to say that he does not 'consider ... authoritarian regimes which have begun to move in the direction of growth and welfare ... normatively inferior to developed democratic ones' (Almond 1973: 268). Thus symbols representative of a precarious historical period, such as 'growth,' 'development,' 'welfare,' and episodical institutional entities are themselves transformed into evaluative criteria of political reality at large.[20] In this vein a new area of study, 'institution-building,' addressed to the 'third world' by Western academic headquarters, has been lately conceived on the assumption that such an endeavor requires only operational expertise.

It is not surprising to realize that formal political theory finds itself undermined by contemporary events which lay bare the historical precariousness of its standards. Its notions of 'political order' and 'political development,' stripped of explicit substantive and ethical dimensions, have proven to be so theoretically inconsistent that the ideological character of such a discipline can hardly escape attention. This critical situation is not at all peculiar to political science alone; it characterizes the entire domain of formal social science, as long as it persists as a facet of the serialist adventist ideology.

Apparently, formal political science has dismissed the notion of a good polity as a legitimate theoretical concern. Nevertheless, and precisely when its comparative criteria are considered, formal theory covertly implies that the good and best are represented by symbols like 'growth,' 'welfare,' 'industrialization,' and the institutional apparatus which enables societies to reach these goals. For instance, in the cybernetical language nowadays in fashion the polity is defined as a mechanomorphic system and the statesman as its operator (Deutsch 1966: 182–5). In this view one is bound to admit that the accuracy of political knowledge is mainly dependent on the quality and quantity of information available. And if, as it has been estimated, the amount of information now needed for accurate political knowledge would require millions of IBM standard cards,[21] practically speaking, computers rather than human beings should be given the decisive role in steering contemporary polities.

It may be largely true that the steering of advanced industrial polities may actually rely on quantitative criteria (Deutsch 1966) rather than ethical normative criteria. However, to attribute a paradigmatic character to the condition of these polities is tantamount to legitimizing the blind course of human history. Such an information approach (Deutsch 1966: 145–62) crudely freezes the present world power configuration, and ultimately reads the dichotomy between 'developed' and 'underdeveloped' as an ethical, historical verdict, when in fact it exists in people's feelings of relative contentment and deprivation rather than in the concrete possibilities of their contexts.

CONCLUSION

All extant organization theory necessarily presupposes a social science of the same epistemological nature. The counterpart of current organization theory is formal social science. The counterpart of the new science of organizations is substantive social science. In this chapter a brief characterization of these two types of social science was presented. I should point out, however, that although in different passages of this chapter classical theory was utilized to illustrate the distinction, the fundamental assumptions of a substantive theory of human associated life are drawn from the exercise of a sense of reality common to all individuals at all times and everywhere. There is a legacy of human thought which transcends classical theory in the strictest sense and which is active and operative in the minds of several contemporary scholars sensitive to the precarious character of the modern age. Such sensitivity, however, is missing or latent in the representatives of formal social sciences. The mood characteristic of the modern age reflects an unarticulated assumption about human nature. This mood implies that human nature itself is all but historical. It is therefore evident that formal social science can never reach the level of a truly critical theory. Indeed neither history nor society can criticize itself because the yardstick for such evaluation does not lie in any of their episodical configurations. Rather the yardstick is a component of the basic structure of human nature, which actualizes itself differently in different cultures and periods. History is a continuous, intelligible symposium in which all generations and societies understand each other. But it is not history itself that allows us to be intelligible and intelligent. Rather it is reason, in the substantive sense, which enables human beings to understand the historical varieties of the human predicament.

There is a vicious circle linking formal social science and the modern mood whose lure continues to be so effective that most people do indeed make decisions about themselves and social affairs primarily according to the assumptions characteristic of such a mood. The obfuscation of common sense by the modern mood constitutes the essence of what I propose to call the behavioral syndrome, the analytical discussion of which is to be undertaken in the next chapter. The identification of such a syndrome as a deformation, rather than a paradigm, of human nature is essential if we are to understand the psychological underpinnings of the old and the new organization theory.

3

The Behavioral Syndrome

Current organizational theory fails to provide an accurate understanding of the complexity of social systems analysis and design. This failure largely results from its psychological underpinnings. Thus the development of a new science of organizations requires an analytical explanation of this psychological foundation. Before we begin this task, several preliminary considerations are in order.

First, organizations are cognitive systems; organizational members generally internalize these systems and thus unknowingly become unconscious thinkers. But organizational thinking may even become conscious and systematic when it is articulated with fundamentalist overtones. This kind of thinking is characteristic of 'theorists' who articulate the cognitive system inherent in a particular type of organization as a normative cognitive system in general.

Most of what is usually called organization theory lacks scientific rigor and is, rather, disguised tautology or at best refined organizational thinking. Such thinking accepts at face value criteria inherent in organizations. It is a by-product of organizational processes themselves. It considers as normal and natural organizational requirements as they are found eventually superimposed upon human conduct at large. In contradistinction, a true scientific organization theory has criteria of its own, i.e., criteria not necessarily identical with those of social and organizational effectiveness. A scientific organization theory is not predicated upon cognitive systems inherent in any existing type of organization. Rather, it checks organizations against the understanding of conduct generally proper to human beings, taking account of both substantive and functional requisites.

Second, a distinction between behavior and action is here proposed in order to clarify the psychological reductionism of extant organization theory. Behavior is a mode of conduct predicated upon functional rationality or utilitarian reckoning of consequences, a capacity, as Hobbes correctly pointed out, that the human individual has in common with other animals. Its cardinal category is expedience. Behavior is accordingly void of generally valid ethical content. It is a mecanomorphic type of conduct dictated by external imperatives. It can be appraised as functional or effective. It is completely included in a world solely determined by efficient causes.

In contradistinction, action is proper to an agent who deliberates about things because he is conscious of their intrinsic ends. By acknowledging such ends, action is an ethical mode of conduct. Social and organizational effectiveness is an incidental, not cardinal dimension of human action. Human beings are bound to act, to make decisions and choices, because final, not only efficient causes, have a bearing upon the world at large. Thus action is predicated upon utilitarian reckoning of consequences only by accident at best.

Before turning to an identification and examination of the psychological underpinnings of extant organization theory, additional preliminary considerations are in order. The next point to be clarified is the linguistic origin of the term 'behavior' and its relation to the pervasive mood of the behavioral syndrome. Subsequently it will be pointed out that the behavioral syndrome emerged as a consequence of an unprecedented historical effort to mold a social order according to economizing criteria. Finally some consideration will be given to the concepts of the good man and good society insofar as they bear upon understanding the concept of the behavioral syndrome.

It is not an accident that in the Western word *behavior* has only recently become a *lingua franca* denoting the patterns of adult interpersonal relationships. The word was never used in scholarly language before the fifteenth century. According to lexicographers, it began to gain linguistic currency around 1490 and denoted conformity to commands and manners dictated by external conveniences. Even today the word has not lost its original meaning. Behavior continues to be a category acknowledging conformity, a fact that is generally overlooked because conformity to socially given criteria of gregariousness has been transformed into the standards of human morality in general. Men and women no longer live in communities where a substantive common sense determines the course of their actions. They belong instead to societies in which they do little more than respond to organized inducements. The individual has become a behaving creature.

The behavioral syndrome is a socially conditioned mood affecting individuals' lives when they confuse the rules and norms of operation peculiar to episodical social systems with rules and norms of their conduct at large.

The behavioral syndrome, i.e., the obfuscation of the individual's sense of standards generally proper to human conduct, has become a basic characteristic of contemporary industrial societies.

Such societies are the culmination of a historical experiment, now about three centuries old, which attempts to create an unprecedented type of human associated life ordained and sanctioned by the self-regulative processes of the market. The experiment has succeeded, but certainly too well. Not only have the market and its utilitarian ethos become overarching historical and social forces, in their large-scale institutionalized forms they have also proven to be highly expedient for escalating and exploiting nature's processes and for maximizing human inventiveness and productive capabilities. However, throughout this experiment the individual has deceptively gained material improvement in his existence and paid for it with the loss of a personal sense of self-orientation. The exemption of the market from political regulation has given rise to a type of human associated life ordered solely by the interplay of the individual interests (for self-preservation), i.e., a society in which sheer calculus of consequences substitutes for man's common sense.

As I explained in chapter two, a formal social science which is predicated upon the notion of the human being as a behaving creature is a misnomer. This so-called science equates human nature with the characteristics of a certain type of society which is itself a mere accident in history. This 'science' deals with socialization, acculturation, and motivation as if the standards of the good were inherent in such a society.[1] Rather, it must be understood that all societies are less than good; only the human being eventually deserves to be characterized as good. The good man in turn is never a completely socialized being; rather he is an actor under tension, yielding to, or resisting social stimuli on the grounds of his ethical sense. Actually, a society can never approach 'goodness' because of the unregulated processes of the market. Up to a certain degree such a condition is only reachable through the deliberations of its members at seeking to design the good society.

Concluding these preliminary considerations, I turn now to an analytical discussion of four main features of the behavioral syndrome: (1) the fluidity of the self, (2) perspectivism, (3) formalism, and (4) operationalism. Connections between each of these features and the market mind-set will be indicated.

THE FLUIDITY OF THE SELF

The 'fluid self' is an expression used by Arnold Hauser in his study of mannerism, the early stage of modern art. Hauser points out that the 'fluid self' and other characteristics of mannerist artists anticipated the mood which in later centuries became a psychological syndrome of capitalist societies. Hauser singles out Montaigne as a typical mannerist writer and as one of the best illustrations of the fluid self. He interprets the French philosopher as claiming that valuation of things has no permanent ground and that 'nothing is either good or bad in itself' (Hauser 1965: 49). Created by the individual, values are not 'everlasting, immutable, and unequivocal ... Human nature [is] infirm and inconstant, in a state of perpetual flux, hovering between different states, inclinations, moods, for it is in continual transition ... and its true nature lies not in permanence but in change' (Hauser 1965: 49). Thus, significantly, Montaigne in his *Essays* writes: 'I do not portray being; I portray passing ... If my mind could gain a firm footing, I would not make essays, I would make decisions, but it is always in apprenticeship and on trial' (Montaigne 1975: 611).

I submit that the fluidity of the self cannot be fully explained without linking this phenomenon to the mode of representation through which capitalist societies legitimize themselves. Implied in contrasting modes of representation characteristic of medieval, as well as ancient and a number of contemporary non-Western societies, is the assumption that the universe at large constitutes a coherent order and that the human community itself is part of this order. Each of these societies envisions itself as a continuous but precarious re-enactment of the cosmic order in a world of history and change. What the problem of representation poses to these societies is the truth of their existence. The source of their self-interpretation is a metahistorical paradigm which offers a referent appropriate as a normative framework for human conduct at large.[2] As mundane history and changes in circumstances continuously undermine the individual's consciousness of this paradigm, these societies period-ically undertake ceremonies of self-purification and restoration of the common sense of metahistorical grounds. In these societies individuals find a consistent basis to develop their identities and there are patterns for many roles and callings, but these patterns provide the means of expressing individual identity.

In modern societies representation is a purely sociomorphic process; it is no longer a legitimization of the truth of communal existence on the metahistorical grounds. It is, rather, a requirement for negotiated peacemaking between individuals, to enable them to maximize their individual interests.[3] Modern society does not acknowledge itself as a replica of a larger cosmos, but as a contract between humans written large.[4] Human conduct is supposed to comply with utilitarian criteria and this compliance, in turn, fosters the fluidity of the self. Indeed modern man is a fluid reckoning creature who essentially behaves according to objective rules of expedience.

It is no accident that Hobbes, who is the most authoritative source of the modern concept of representation, conceives of the individual as a rule follower. For him, good and bad are mere names whose meanings are conventionally set. In this view of representation, 'fairness' substitutes for truth. Society is a system of rules of a certain kind. If the individual agrees to participate in it, he acknowledges that his conduct is bound within the orbit of a contract. The good citizen abides by externally derived prescriptions. True action is alien to his transactions with others. He can only behave.

The phenomenon characterized by Hauser as the fluidity of the self is peculiar to modern society and constitutes one of the main facets of the behavioral syndrome. In modern society the state of mundane everyday affairs is not supposed to be checked against a paradigm of cosmic order. In such a situation of metahistorical void the individual lacks the 'firm footing' necessary for his identity to unfold. Rather he is compelled to cope with processes and changes which are derivative from a self-induced indefinite movement of the social aggregate. Modern man is the dupe of a misplaced faith. As the Bible has it, to believers faith is the hope for things which although unseen are continuously affecting the universe and bestowing meaning upon the course of events. But having no metahistorical and metasocial roots the secularized modern mind has transposed rather than eliminated its articles of faith; it now believes in society's invisible hand.

A certain epistemology is to be drawn from this condition, according to which processes and changes are to be explained as exclusively activated by efficient causes. This is, incidentally, the epistemology on which conventional social science relies. In the prevailing language of conventional social science, expressions like 'process-orientation' and 'change-orientation' entail the misplaced faith in the absolute transitoriness of things. This understanding of processes and changes is defective and one-sided for logical reasons which I will indicate, drawing from the philosophical investigations of Alfred North Whitehead.

One of the many merits of Whitehead's philosophy of organism is the elucidation of the notion of process. He acknowledges as valid Heraclitus's intuition that 'all things flow.' But he deems it more proper to speak of 'the flux of things,' since all flux is a flux of something (Whitehead 1969: 240). The 'flux of things' is the accomplishment of their inherent patterns, and therefore it results altogether from efficient and final causes. The 'flux of things' is objectively conditioned by the data constitutive of the given world and also by the private experience of their aims. There is no indefinite flux from nothing to nothing. To be precise, things become selectively, finitely. Unselective, indefinite becoming is inconceivable. Things are finite, epochal processes. They perish as processes, although they are everlasting as patterns. In a nutshell, Whitehead renders his idea of flux as follows:

There are two principles inherent in the very nature of things, recurring in some particular embodiments whatever field we explore – the spirit of change, and the spirit of conservation. There can be nothing real without both. Mere change without conservation is a passage from nothing to nothing. Mere conservation without change cannot conserve. For after all, there is a flux of circumstance, and freshness of being evaporates under mere repetition. The character of existent reality is composed of organisms enduring through the flux of things. (Whitehead 1967: 201)

In essence Whitehead elucidates that apart from being, passing is inconceivable.

One might ask what in modern history engendered the widespread feeling of the permanent transitoriness of all things, so well articulated by Montaigne. A part of the answer is given by the concept of nature characteristic of modern science since the seventeenth century. Modern science reads nature ultimately as bits of matter in motion and represents values as adventitious to nature. In the domain of everyday life this view is filtered down as a feeling of an aimless, permanent transitoriness of things. Reduced to bits of matter in motion things are conceived as having an infinite tolerance to change. If values and aims cannot be considered as inherent in things themselves, they are bound to be enchained in a world in infinite progression. In this world there is no becoming, since the becoming of things cannot be assessed apart from the assumption that they are endowed with private experiences of reality, or ends of their own.

The question also admits another partial answer. The feeling of the aimless, permanent transitoriness of things is a consequence of the uncritical internalization by the individual of the self-representation of modern society. This society defines itself as a precarious contract between utility-maximizing individuals in the pursuit of their happiness, understood as a pursuit of satisfaction of an endless succession of desires. There is no meaning for this striving beyond the social confines. Since, given its competitive character, the social world at large becomes unfamiliar to him, the individual tries to overcome his estrangement either by annulling himself through passive conformity to roles prevailing here and there, or by withdrawing into himself and thus asserting his all too conscious identity. But since the ordering center of his life is nowhere, his identity is his creation. This cultivation of the self relapses into narcissism.[5] A psychology, itself oblivious to whatever may transcend the social inducements bearing upon the human psyche, comes to the help of the individual. Ours is an age of psychological refurbishing. In psychological clinics the estranged individual is encouraged to undertake the search for his self. It is, however, questionable whether this search can ever succeed in a world ordered according to contractual rules of a social aggregation of competitive interests. When the human predicament is assumed to be only social, the fluidity of the self is inevitable.

PERSPECTIVISM

In conforming in a fluid manner to society understood as a system of contracted rules, the individual is induced to realize that both his and other's conduct is affected by a perspective. He becomes a perspectivist. Certainly perspective is always an ingredient of human conduct in any society. But only in modern society does the individual become conscious of this fact. Such a society engenders the peculiar type of conduct which deserves to be denoted as behavior. To behave well, then, the individual has solely to take account of external conveniences, others' viewpoints, and purposes at stake.

In discussing the increasing salience of perspectivism as a fundamental aspect of the psychological foundation of prevailing organization theory, it will be helpful to give some historical background to the term. Perspective, as a dimension of human expression, first became a technical term in the domain of painting. In fact all painting styles have been characterized by a certain perspective. But only in the late period of the Middle Ages did perspective begin to constitute an object of the painter's attention. Giotto (1276?–1337?) already assumes that what the artist offers on a canvas is not a copy of nature, but nature according to the eyes of the painter. Later on, Petrarch (1304–74) would echo Giotto in his dictum: 'Everyone should write his own style.' But it was Leon Battista Alberti (1404–72) who envisioned the laws of perspective as an object of formal scientific investigation. Subsequently, in the wake of the commercial and industrial revolution, perspective was increasingly to become a systematic category of artistic work, as well as a characteristic of human conduct at large.

In the sixteenth century an artistic market flourished in Italy. What persons of taste buy in such a market is very much the personal view of the artists. The personal element becomes the trademark of the works of art. Appreciation of them requires a certain initiation in the peculiar manners of the artists. Historians cogently stress *mannerism* as the basic characteristic of art in this century when the *connoisseur* finds for the first time an opportunity to make a living in the artistic centers of Italy. He is a merchant and broker between the artists and their affectionates, and the artists already realize that they work for a market. The concept of intellectual property, unknown in the Middle Ages, is now acknowledged. The artist is an entrepreneur entitled to claim property rights upon his works which he may or may not sell to the public according to the market price.[6]

Yet perspectivism at this time is not confined to the artistic milieux. It is a trait of the everyday life of an increasing number of people involved in activities propitiated by the emerging market system. The market is indeed the underlying force which engenders the perspectivist view of human associated life. One might say that Machiavelli finds in the conditions prevailing in his time the inspiration to elaborate his political theory.[7]

Perspectivism pervades Machiavelli's thought. An example of this is the analogy used by Machiavelli in the dedication of *The Prince* to Lorenzo de Medici. The dedication itself is a device of expedience, serving his intent of drawing personal advantages from his flattering of the prince. But what must be stressed is Machiavelli's characterization of the correct way to study the art of ruling. He compares the students of politics with 'those who draw maps of countries.' He explains: they 'put themselves low down on the plain to observe the nature of mountains and of places high above, and to observe that of low places put themselves high upon mountain tops, so likewise, in order to discern clearly the people's nature, the observer must be a prince, and to discern clearly that of princes, he must be one of the populace' (Machiavelli 1965: 10–11). Machiavelli resorts to this perspectivist metaphor in order to state that the study of politics requires an integration of the viewpoints both of the prince and of the people. To use Mannheim's terminology, Machiavelli is already a full-fledged 'relationalist.' But his 'relationalism' is unconcerned with truth, even in a relative sense. It is essentially concerned with expedience. The prince needs to be instructed about the perspective of the ruler in order to preserve and enhance his assets. He needs to understand the perspective of the 'ordinary citizen' in order to deceive him. The prince has to be sensitive to scenic imperatives, i.e., to be virtuous by simulation, and able to induce the citizens to be good through the 'wise' exercise of cruelty.

Machiavelli systematically distorts theoretical language by stripping it of any ethical substance, a practice in which Hobbes later will excel. For instance, with Machiavelli prudence gains an unheard-of connotation. Machiavelli's idea of prudence is void of ethical content. Prudence is sheer calculus of consequences, a skill at the service of interests. He is the founder of an 'interest theory of politics' (Wolin 1960: 233) in which 'cruelty,' 'trick,' 'deceiving,' 'miserliness,' 'war,' and 'mass killing' are sanctioned as legitimate expressions of human conduct. He thus praises deeds which are deplorable to common sense. The prince should not feel as his duty the practice of qualities 'considered as good' for they may bring his 'destruction.' There are qualities 'that look like vices, yet – if he practices them – they will bring him safety and well-being' (Machiavelli 1965: 59). To be sure 'everyone,' he says, 'will admit that it would be most laudable for a prince to exhibit [the] qualities ... considered good ... But no ruler can possess or fully practice them, on account of human conditions that do not permit it' (58). Machiavelli's teaching implies that not only princes but common individuals as well are entitled to suspend ethical standards of good action in the pursuit of their interests. He is indeed one of the first modern thinkers who understood the motivational patterns inherent in a market-centered society. Those patterns at large, and perspectivism in particular, have become normative standards of human conduct.

FORMALISM

Formalism is a third aspect of the psychological foundations informing current organization theory. It is a characteristic of human conduct that has become externally oriented. The term is usually employed by historians of art, including Arnold Hauser, to describe a particular psychological characteristic of Western society in its early capitalist period. Formalism is still useful today as an explanatory category of human conduct. Indeed, formalism has become a normal feature of everyday life in market-centered societies. In these societies abiding by the rules substitutes for a concern for substantive ethical standards. Exposed to a world pervaded by ethical relativism, the ego-centered individual feels estranged, and in order to overcome such estrangement he indulges in formalistic kinds of behavior, i.e., he conforms to external imperatives of getting things done. He becomes a 'mannerist.' To be sure, 'mannerism' is the psychological disposition required by a type of politics divorced from concern with the good, a kind of economics uniquely concerned with value-in-exchange, and a science in general essentially defined by method and operational procedures.

Behavior is a manifestation of mannerism. It is entirely subsumed under the incidental criteria of the public arena. Its meaning exhausts itself in its appearance to others. Its reward is its outward recognition as adequate, correct, or fair. Its subject is not a consistent self, but a fluid creature ready to perform expedient roles.

An insightful view into the nature of the behavioral syndrome may be drawn from a careful reading of two mannerist documents, *The Courtier* by Baldesar Castiglione, published in 1528, and *The Theory of Moral Sentiments* by Adam Smith, published in 1756. Since these documents are especially revealing not only with respect to the behavioral syndrome in general but with regard to formalism in particular, they will be considered in the following paragraphs.

Courts have always existed in some form throughout history. But the court as it appeared in Italy, Spain, France, and other European countries in the early periods of capitalism was a peculiar historical phenomenon. More than ever before, the court then was the overarching center of associated life. It was the arena in which the significant actors of public life regularly met. Nothing in the religious, political, economic, military, artistic, and other domains of public life gained a normative character at large without first being screened by the court. Since the court had a decisive bearing upon everyday human affairs, the manners prevailing among those admitted to its circle became the norms of good human conduct in general. By identifying himself with those manners without critical detachment, Castiglione transforms historically precarious criteria into criteria of good human conduct. It is just such an identification which is the essential feature of behavioralism.

It is to be noted that Castiglione places himself in the company of

Cicero, who also is an author of several tracts on norms of human conduct. But if one reads carefully Cicero's *De Officiis*, for instance one realizes that the Roman writer never succumbed to the episodical lure of the purely courtly. Cicero is basically concerned with what is generally good beyond any particular episodical circumstances. Thus he writes: 'Even though [the] moral goodness [of a human conduct] be not generally ennobled, [it] is still worthy of honor; and by its own nature ... it merits praise, even though it be praised by none' (Cicero 1975: 17).

In contradistinction, Castiglione is basically concerned with social approbation. Accordingly he describes courtly behavior as a general paradigm of human conduct. His book is not a treatise in education as are Plato's *Republic* and Cicero's *De Officiis*, authors Castiglione claims to be imitating, but presents a methodology required for gaining good reputation, a treatise on 'management of impressions'. For instance, talking about 'nonchalance' as a characteristic of commendable conduct, one personage of *The Courtier* says that 'apart from being the real source of grace [nonchalance] brings with it another advantage; for whatever action it accompanies, no matter how trivial it is, it not only reveals the skill of the person doing it, but also very often causes it to be considered far greater than it really is. This is because it makes the onlookers believe that a man who performs well with so much facility must possess even greater skill than he does, and that if he took great pains and effort he would perform even better' (Castiglione 1976: 70).

All through his book Castiglione suggests that the only reward of good conduct is public praise. For him there is no good conduct in its own right. Thus one of his characters advises the courtier to be 'mindful of where and in whose presence he is' and to 'feed the eyes of those who are looking on' (Castiglione 1976: 116). In a later period of Western society the public arena will transcend the court and will become 'society' itself. The rules of prevailing social behavior will become rules of good conduct in general.

At this juncture of the analysis, elucidation of the nature of formalism can be drawn from Aristotle's statement that 'a man is meant for political association, in a higher degree than bees or other gregarious animals can ever associate' (Aristotle *Politics*, I: 1253a, para. 10). Consistently with this he also points out that a 'good man' is not necessarily a good citizen. What is emphasized by Aristotle is that the good man is primarily guided by what is here qualified as substantive reason, which is common to all men, everywhen and anywhere, and is not to be considered coincidental with particular standards of any given society. In his *The Theory of Moral Sentiments*, however, Adam Smith twists the age-old meaning of reason in order to make the term consonant with economizing criteria, and replaces it with the individual's sentiment of gregariousness. He writes:

Though reason is undoubtedly the source of the general rules of morality, and of all the moral judgements which we form by means of them, it is altogether absurd and unintelligible to suppose that the first perceptions of right and wrong can be derived from reason ... These first perceptions, as well as all other experiments upon which any general rules are founded, cannot be the object of reason, but of immediate sense and feeling ... But reason cannot render any particular object either agreeable or disagreeable to the mind for its own sake. (Smith, 1976: 506)

In Adam Smith's view, as well as in the view of all who claim that morality is congruent with sociality itself, the individual is left with no metasocial 'firm footing' to determine responsibly the ethical character of his conduct. Man does not properly act, but behaves, i.e., is bound to abide by the eventual rules of social approbation. Therefore education is not meant to develop the individual's potential to become a 'good man' in Aristotle's sense. 'The great secret of education,' says Smith, 'is to direct vanity to proper objects' (Smith 1976: 417). The correctness of human conduct lies in its very form, not in its intrinsic content. The individual should place himself before a 'looking glass by which, [he] can ... with the eyes of other people, scrutinize the propriety of [his] own conduct' (Smith 1976: 206). There is thus no essential disagreement between Castiglione and Adam Smith in regard to the method suited to the determination of norms of human conduct in general. The only difference between them is of an episodical nature. For Castiglione the 'looking glass' for the individual is the court. For Adam Smith it is 'society.' Adam Smith's *Theory* is, therefore, equivalent to Castiglione's *The Book of the Courtier* in that Smith's *Theory* succumbs to the same lure of the episodical, whose model is transformed into a standard of general human conduct.

The legitimization of episodical forms of human conduct according to their precarious inherent principles continues until today to be a basic postulate of 'objective,' 'value-free' behavioral science. It is therefore understandable that today's practitioners of behavioral science essentially see themselves as students of processes. Form, not substance, is what matters. For these practitioners the behavioral syndrome is a given, and to question it is pointless.

OPERATIONALISM

As currently understood, operationalism tries to respond to the following question: how to assess the cognitive character of a statement? There are two basic answers to this question. One type of answer admits that there are several kinds of knowledge (such as metaphysics, ethics, and physics), each of which requires specific procedures of verification. There are those, however, who claim that only the procedures inherent in the method of a mathematized natural science are adequate for validation and verification of knowledge. The latter answer constitutes the thrust of what is here labelled operationalism.

It seems prudent, however, to qualify operationalism thus understood as positivist operationalism, lest one interpret this section as suggesting that criteria of rigorous thought are irrelevant in all disciplinary fields but physics. On the contrary, wherever articulation of thought does not meet criteria of rigorousness there is not scholarship. That is why it is not easy for one to engage in conversation in the domains of metaphysics, ethics, or aesthetics without mastering their specific patterns of thinking and elucidation of matters. For instance, what is conceptualized as form in metaphysics, goodness in ethics, or beauty in aesthetics is not given to human perception in the same fashion as size, shape, extension, and number of objects. Nevertheless one can legitimately argue that to think of things as actualizing forms is a requirement for the apprehension of their concreteness. By the same token, judgments that an individual is good, and a work of art beautiful, imply that goodness and beauty are real objects of a certain kind, not directly apprehended by immediate sensory perception.

One might consider as representative of positivist operationalism P.W. Bridgman's contention that 'a concept is nothing more than a set of operations.'[8] G. Lundberg expresses his idea of operationalism by asserting that the 'recipe for a cake defines a cake.'[9] Possibly not all positivist operationalists would literally agree with Bridgman and Lundberg, but in general they all share their doctrine that only what can be physically measured or assessed deserves to be considered as knowledge. This is one of the reasons why in the operationalist milieux the word metaphysics is fraught with pejorative connotations. To say that a statement is metaphysical is equivalent to dismissing it as meaningless.

Positivist operationalism may be considered as a feature of the behavioral syndrome. In other words, whoever subscribes to positivist operationalism is caught within the confines of a peculiar psychological bent. In analyzing the psychology of operationalism, two of its main characteristics must be emphasized.

First, positivist operationalism is pervaded by a control orientation toward the world and thus induces the researcher to focus upon its manageable aspects. This characteristic arises from philosophical and psychological referents.

Philosophically and, as a matter of fact, metaphysically, positivist operationalism reflects the view of the universe inherent in classical physics. For instance, Galileo intimated that what is real in the world can only be thought of as extension, space, mass, motion, and solidity. Accordingly, the conceptual apparatus to approach reality is necessarily to be drawn from mathematics. Indeed, modern mathematics takes account in nature only of those aspects which can be quantitatively expressed. In general, classical physicists consider these aspects as the only qualities, and claim as secondary (i.e., figments of imagination) whatever other qualities the mind conceives. Thus concepts of high-grade abstract order such as those which constitute modern mathematics postulate what things in the world are to be understood as real or unreal. This substitution of the abstract for the concrete is precisely what White-head correctly identifies as the 'fallacy of misplaced concreteness' (White-head, 1967: 51).

Hobbes accepted Galileo's doctrine and accordingly developed his conception of 'civil philosophy,' an expression which encompasses what is today known as political and social science. Thus he states that feelings such as 'love,' 'benevolence,' 'hope,' and 'aversion' (mere 'motions of the mind' induced by external influences), as well as human conduct in general 'are to be considered after physics' (Hobbes 1839: 72). Hobbes would claim that social science is necessarily social physics of a certain kind and that the problem of order in human affairs admits only a mechanical solution. Since the notions of good and evil and all virtues and affections belonging to the domain of ethics assume the character of secondary qualities, designing a good society is equivalent to designing a mechanical system in which the individuals are geared, by external inducements, to abide by rules of conduct needed to maintain the stability of such systems.

Hobbes's science of man and society, modelled according to classical physics, although in mitigated forms, is still influential today among students and practitioners of behavioral science, operational research, and systems analysis and design of a certain kind. These currents are in fact outliving the decline of the type of physical science from which they are derived. Indeed today's physics tends to work with imageless concepts which cannot be articulated as recipes, since it denies to sensory perception a paramount role in theory building. Having restated the concept of becoming into physical reality, theory building in today's physics is, rather, art and obeys aesthetical rules.[10]

Psychologically, the control orientation toward the world has been inherent in positivist operationalism since its historical inception in the seventeenth century. Its founders in physics are characters like Galileo who were reacting against the dominant and dogmatic contemplative mood of medieval thinkers. Modern thinkers wanted the practical world to be the proper object of scientific inquiry. Galileo's refutation of Aristotle's doctrine of the fall of bodies, through his experiment at the Tower of Pisa, illustrates a case where validation of knowledge requires more than syllogistic reasoning.

An interest in dealing with practical problems of this world is at the root of operationalism. This interest was made explicit by Francis Bacon in his *Novum Organon* where he stated that 'knowledge is power.' Consistent with this orientation is Bacon's assertion that 'what in operation is most useful, that in knowledge is most true' (Bacon 1968: 122). It is in this sense that what perverts operationalism is its identification of the 'useful' with the 'true.' Utility is a notion fraught with ethical ambiguity. In itself, what is useful may serve to do both the ethically sound and the ethically wrong in the social domain, and thus the role of operationalism in social science should be ethically qualified. This is precisely what Hobbes and conventional social scientists in general fail to do. They have stripped utility of its ethically ambiguous character by legitimizing as generally normative what is useful for the social system to control its human participants. Once more the affinity between operationalism and the behavioral syndrome is evident.

Positivist operationalism bears upon the behavioral syndrome in another way. It refuses to grant final causes any role in the explanation of the physical and social world. It implies that things are only outcomes of efficient causes, the entire world being a mechanical chain of antecedents and consequents. This assumption is a systematic component of the doctrine of Galileo, Newton, Laplace, and all those who envision social science as an extension of classical physical science. Since final cause is an expression which rarely occurs in current technical language, no one in the context of this analysis can serve as a better example than Hobbes to make the reader sensitive to the question here at stake.

Hobbes expresses the idea of causation implied by positivist operationalism as follows: 'A final cause has no place but in such things as have sense and will; and this also I shall prove ... to be efficient cause' (Hobbes 1839: 132). No wonder that in order to be consistent with this doctrine Hobbes was led to define reason as nothing but calculus of consequences in the mechanic sense. The qualification 'in the mechanic sense' is needed here because in the process of their actualization things are indeed under the bearing of some sort of antecedents and consequents. In their actualizing process things find data which are their efficient causes. But these data are not the only determining agent of the process of things, as Hobbes and positivist operationalists in general claim. For instance, Hobbes boldly states that 'nothing takes beginning from itself, but from the action of some other agent without itself' (Hobbes 1840: 274). He conceives the universe as a mechanical order, the understanding of which requires a reasoning of mathematical kind, a calculus essentially consisting of addition or subtraction. Findings of contemporary science show that this conception of causation is untenable. For instance, certainty in predicting the process of things is admitted as theoretically possible in the mechanist idea of causation, whereas Heisenberg's principle of uncertainty, empirically substantiated, implies that things have ends of their own which endow them with some quantum of self-determination. They are indeed affected by antecedents in the sense that, not existing in abstraction, they have to appropriate the data given in the world, but such appropriation is not to be explained as a passive accommodation to external circumstances; it is a selective process of exclusion and inclusion of data according to their private aims. In Whitehead's language, things are continuously prehending data in the accomplishing of their inherent patterns. Thus contemporary science reinstates final cause in the physical and social domain.

Hobbes correctly understood that one could not subscribe to positivist operationalism without reducing man to a mecanomorphic kind of entity. Thus Hobbes consciously equates liberty with necessity. Accordingly, it should not be attributed to humans alone, but to all bodies. 'Liberty,' he says, 'is the absence of all the impediments to action ... as for example, the water is said to descend *freely*, or to have *liberty* to descend by the channel of the river, because there is no impediment that way, but not across, because the banks are impediments' (Hobbes 1840: 273–4). Therefore man never acts properly, but always yields to external inducements, for his 'will ... and each propension of man during his deliberation, is as much necessitated, and depends on a sufficient cause [which in Hobbes is the same as efficient cause] as anything whatsoever' (Hobbes 1840: 247).

Even God does not escape the bearing of mechanical necessity. 'God,' he says, 'does not all things that he can do if he will; but that he can *will* that which he has not *willed* from all eternity I deny' (Hobbes 1841: 246). In the terminology of the present analysis, this is tantamount to saying that God and human beings do not act. They can only behave. The world unfolds according to a blueprint established once, and for all eternity. There is no creativity in Hobbes's mecanomorphic universe.

CONCLUSION

Striking as the basic features of the behavioral syndrome might appear, one should realize that they are not affecting people's lives remotely. In fact they constitute the unarticulated creed of institutions and organizations operative within the confines of the market-centered society.

In order to cope with the challenges of such a society, most of its members internalize the behavioral syndrome and its cognitive patterns. Such an internalization usually takes place unnoticed by the individual, and thus the behavioral syndrome becomes second nature. Standard administrative scholarship, itself largely confined within preconceptions and assumptions of human beings as fluid selves, and by perspectivism, formalism, and operationalism, can not help the individual to overcome this situation. The next two chapters will focus upon two outcomes resulting from the captivity of standard administrative scholarship, namely, misplacement of concepts and cognitive politics.

4

Misplacement of Concepts and Organization Theory

The field of organization theory has been so promiscuously receptive to influences from so many different areas of knowledge that it now seems to have lost a sense of its specific assignment. Although cross-disciplinary relations are in general positive and even necessary to creativity, it is time for a serious appraisal of the state of the field, lest it become a mere hodgepodge of theoretical ramblings, lacking both force and direction. Any discipline must have a modicum of intolerance in its transactions with other disciplines, otherwise it will lose its reason to exist. To have identity and character is in a sense to be intolerant.

In this chapter I contend that the extrapolative process, which I call misplacement of concepts, is rendering organization theory characterless; it will be crippled if it continues to indulge in the practice of unqualified borrowing from other disciplines, theories, models, and concepts alien to its specific task.

BASIC FEATURES OF THEORY BUILDING

Theory building in the organizational field has most frequently occured as a result of (a) primary original creation, (b) serendipity, and (c) displacement of concepts.

It is not my intention to inquire into the complexities of conceptual creativity. Suffice it to say that a concept results from a primary act of creation when no antecedent of it is apparent, when it has been drawn from nothing but the direct personal transaction between the thinker's mind and the peculiar features of the issue or problem under attention. Thus, if we believe Cassirer, Montesquieu 'is the first thinker to grasp and to express clearly the concept of "ideal types"' (Cassirer 1951: 210). With the systematic characteristics later attributed to it by Max Weber, this concept brought an unprecedented insight into the nature and meaning of theory building itself. Concept formation, however, usually results from serendipity and displacement of concepts, true original conceptual creation being more rare than is ordinarily admitted.

As explained by Robert Merton, serendipity takes place when 'an unexpected and anomalous finding elicits the investigator's curiosity and [conducts] him along an unpremeditated bypath which [leads] to a fresh hypothesis' (Merton 1967: 108). Serendipity in the organizational field is well illustrated by the so-called Hawthorne Studies. The original purpose of this research was to assess the effect of lighting on the worker's output. In a first attempt no significant relationship was found between the two variables. This unexpected result led the researchers to pursue a thorough investigation of the factors of efficiency, the outcome of which was the discovery that sentiments and informal relations between the employees, as well as their personal needs and social conditions external to the organization, have a systematic influence on productivity. One can say that the efforts at evaluating and discussing the steps and the results of the Hawthorne Studies led Fritz Roethlisberger and William J. Dickson to an incipient formulation of what is now known as the systems approach. (This root of the systems approach has been overlooked by the chroniclers of the field).

The displacement of concepts may provide a fruitful way of gaining insight and may even lead to formulating a 'logic of discovery' (Schon 1963: 54). Such is the endeavor of Donald Schon in his book *Displacement of Concepts*.[1] 'Emergence of concepts,' Schon says, may 'come through the shift of old concepts to new situations' (53). By displacing a concept one tries to understand the unfamiliar in terms of the familiar, or the unknown in terms of the known. Several instances may be cited in the study of politics and administration. One example of displacement of concepts is found in the book *The Principles of Organization* by James D. Mooney and Alan C. Reiley. These authors did explicitly what others like Henri Fayol, Frederick Taylor, and Luther Gulick did implicitly: they deduced from existing historical models systematic guidelines for organizations. Mooney and Reiley, systematically relying on a displacing logic, formulated principles such as the scalar principle, the functional principle, the coordinative principle, and the staff principle, taking as a paradigm of 'organized efficiency' (Mooney and Reiley 1939: 47) the church and the army.

In political science another illustration of displacement of concepts is represented by Karl Deutsch's *The Nerves of Government*. The main thrust of this book is to conceptualize political issues from the 'viewpoint of communications' (Deustch 1966: xxvIII) – that is, by resorting to cybernetic models. In the field of organization theory the equivalent to Deutsch's work is certainly *Social Psychology of Organizations* by Daniel Katz and Robert L. Kahn, in which the main issues and problems of administration are woven into a cybernetic fabric.

DISPLACEMENT BECOMES MISPLACEMENT

While the displacement of concepts can be a valuable, productive, and legitimate means of theory building, it can very easily degenerate into misplacement of concepts. Misplacement of concepts is presently plaguing the field of organization theory. Misplacement of concepts takes place when the extension of a theory model or concept of phenomenon A to phenomenon B does not hold up after a thorough examination because phenomenon B belongs to a peculiar context whose specific characteristics correspond only in limited ways to the context of phenomenon A. One is often in danger of misplacing concepts when engaged in an effort of theory building. The reason why such 'unwarranted extensions of concepts' often happens is offered by Kaplan: 'No two things in the world are wholly alike, so that every analogy, however close, can be pursued too far; on the other hand, no two things are wholly dissimilar, so that there is always an analogy to be drawn, if we choose to do so. The question to be considered in every case is whether or not there is something else to be learned from the analogy if we choose to draw it' (Kaplan 1964: 266).

Thus, in trying to displace a concept, one may be entering into a 'potential intellectual trap' (Nagel 1961: 115)[2] in which the attempt results in the misplacement of a concept.[3]

THE FALLACY OF CORPORATE AUTHENTICITY

Many attempts to displace concepts from other fields into the field of organization theory have given rise to unintended outcomes. In general, concepts are misplaced into organization theory because those who are doing the misplacing do not realize that formal organizations are affected by several forms of sociality which in turn possess different degrees of intensity. For example, taking the degree of intensity of the interface between individuals as a point of reference, Gurvitch (1958: 176) differentiates between mass, community, and communion as forms of sociality. Many writers are led to 'unwarranted' extrapolations, precisely to the extent that they overlook the fact that the third type of sociality – communion – has the least structuring function within formal organizations. A typical illustration of this mistake is what has been called 'authentic organizations.'[4]

Yet 'corporate authenticity' is a contradiction in terms, for authenticity is an intrinsic attribute of the individual: it can never be conquered once and for all. Corporate social existence normally hinders the authentic expression of the individual. Authentic moments of individual life are precisely those in which corporate behaviors are suspended. That is why authenticity is risky. Should it become a corporate steady social state, as some claim, there would be no merit for individuals to be authentic, and the individual would lose the character of a true 'I,' responsible for each individual action. In fact the concept of 'corporate authenticity' ignores the inextricable tension between substantive dimensions of persons and functional requirements of society. Such a concept assumes that a person's actualization can be equivalent to the accomplishment of functional activities. It is true that where communion prevails a high tolerance for authenticity exists, but communion is hardly possible within the confines of formal organizations. Because of its nature a formal organization normally has a low tolerance for individual authenticity, for those types of relationships defined by Martin Buber as I-Thou.

THE MISUNDERSTANDING OF ALIENATION

The idyllic overtone prevailing in many theories current in the organizational field validates Saul Alinsky's contention that 'social scientists often seem to be naïve about the uses and purposes of organizations.'[5] This observation is particularly pertinent to those who claim that it is possible to minimize and even eliminate alienation within the confines of formal organizations.

Contemporary literature on alienation is largely an instance of conceptual misplacement or 'conceptual stretching,' to use Sartori's language. Before Hegel and Marx the thrust of the term 'alienation' was always fundamentally metahistorical and religious. To overcome their misfortune in history, or their estrangement in this world, humans were supposed to come to grips with dimensions of their predicament which were essentially metahistorical and metasocial.[6] Hegel and Marx reduced that estrangement, conceptualized as alienation, to a purely social condition, so that they conceived human de-alienation, redemption, or emancipation not as an event in the life of the individual, regardless of given mundane circumstances, but as the outcome of a certain stage of the historic-social process. Hegel and Marx themselves have misplaced the age-old issue and idea of alienation and/or estrangement.

Yet at least Hegel and Marx were acquainted with the millenial legacy of thought and insights about alienation. They knew that according to this legacy estrangement or alienation was to be considered as a question essentially metaphysical and religious.[7] Their intent was deliberately polemical: to show that human estrangement or alienation could be overcome within the realm of secular history. They implied that prior to them thinkers were focusing upon alienation at a level where it did not belong. Thus they suggested that those earlier thinkers were guilty of misplacing the concept of alienation.

It is hard, however, to find excuses for the current behavioralist literature on alienation. By and large most of its representatives seem to be oblivious to the historical background of the question. Moreover, although some of these contemporary representatives refer to Marx (and rarely to Hegel) as a source, their writings are actually fraught with misinterpretations of Marx's thought, not to mention its antecedents. Apparently their intention is to 'operationalize' Marx's thought. But in carrying out their endeavor they overlook the macrosocial frame in which Marx's concept of alienation solely makes sense. For example, in a widely cited article, Melvin Seeman seeks to make the meaning of alienation 'more researchable' and 'amenable to sharp empirical statement' (Seeman 1972: 46). He portrays alienation as having behavioralist features, such as powerlessness, meaninglessness, normlessness, isolation, and self-estrangement. He justifies his analysis through an out-of-context interpretation of authors like Marx, Max Weber, Durkheim, and Mannheim (gentlemen who *hurleraient de se trouver ensemble*), none of whom would ever accept a behavioralist discussion of the theme. For instance, Seeman says that 'rebellion' closely approximates 'isolation.' If this is indeed true, Marx was an alienated person, which is tantamount to turning Marx against himself. As egophanic a character as Marx certainly was, he would repel with indignation such a suggestion. He thought that his 'rebellion' against the bourgeois society and his 'isolation' within it, were indications that he was anticipating in his lifetime the future historical stage defined by the vanishing of alienation. After all, Marx was a

Hegelian of sorts. Like Hegel, he was convinced that he had deciphered the riddle of history. It is also known that Marx had troubles with his interpreters. He once said: 'Moi, je ne suis pas marxiste.'

To be sure, Seeman is conscious that his understanding of alienation 'departs from Marxian tradition,' but he considers such a 'departure' as 'not radical.' He acknowledges that Marx produced a 'judgement about a state of affairs.' 'My version,' he says, 'refers to the counterpart, in the individual's expectations of that state of affairs' (Seeman 1972: 47). The point is that such 'departure' leads to ludicrous outcomes. This is particularly evident in Robert Blauner's book *Alienation and Freedom*.

Blauner claims to draw upon Marx's theory of alienation. However, in spite of his interesting empirical findings about labor settings in several sectors of American industry, Blauner's research is conceptually precarious, even on Marxian grounds, because the context in which he intends to assess alienation is uncongenial to the global societal context which Marx had in mind. Blauner seems to equate alienation with discontent in job settings. This equivalence amounts to a denaturation of Marx's concept of alienation. As Richard Schacht has correctly pointed out, Marx would not 'hesitate to speak of "alienated labor" even in the absence of actual work dissatisfaction' (Schacht 1970: 164). For Marx, alienation could never be eradicated within the limits of micro-organizations. De-alienation for Marx requires the overall transformation of the global social system itself. What in the existing industrial system makes alienation inevitable is, according to Marx, private ownership of the means of production, an aspect that is systematically overlooked by Blauner. It makes no sense from Marx's perspective to say that 'the printer is almost the prototype of the non-alienated worker in modern industry' (Blauner 1964: 56–7) or that 'the [contemporary] industrial system distributes alienation unevenly among its blue-collar labor force, just as our economic system distributes income unevenly' (p. 5). For Marx, given the intrinsic characteristics of the contemporary industrial societies, alienation is an inevitable feature of everyday life, a total social phenomenon, resistant to any compartmentalized solution. Blauner's research hinges upon a misplacement of Marx's theory of alienation. In fact it represents a misplacement of a misplaced concept, i.e., first, it strips the question of alienation of its metahistorical character; second, it assumes that the question can be solved by micro-organizational means.

When taken at face value by administrative and organizational writers and practitioners, this kind of literature, representative of the behavioralist persuasion, fosters a distorted, if not bizarre, understanding of the interface between formal organizations and their members. For instance, reflecting Seeman's and Blauner's views on alienation, Richard E. Walton claims that the 'roots of worker alienation' can be 'severed' through the 'redesign of the work place' (Walton 1972: 70). Substantiating this statement Walton reports the results of a design implemented in a pet-food plant. Some of the results are thus listed: 'After 18 months, the new plant's fixed overhead rate was 33% lower than in the old plant. Reductions in variable manufacturing costs (e.g., 92% fewer quality rejects and an absenteeism rate 9% below the industry norm) resulted in annual savings of $600,000. The safety record was one of the best in the company and the turnover was far below average. New equipment is responsible for some of these results, but I believe that more than one half of them derive from the innovative human organization' (Walton 1972: 77).

It is the business of practitioners in public and private administration to redesign labor settings, and Walton's article is certainly a research piece of interest to that effect. But to identify de-alienation with job satisfaction and to try measuring it in terms of results like overhead rates, manufacturing costs, annual savings, and safety records is grossly naïve. Further, textbooks help diffuse among students this *naïveté*. For instance, in one generally required textbook, it is said that 'certain types of technology can reduce alienation' (Luthans 1977: 91).[8] In other words, alienation admits of machine-like treatment. This kind of shallow scholarship should be considered inexcusable in graduate schools of business and public administration.

ORGANIZATIONAL HEALTH, A MISNOMER

Another example of a misplaced concept is the notion of organizational health. Warren Bennis claims that a concept of organizational health is needed to overcome the inadequacies of the notion of organizational effectiveness. One can accept Bennis's remark that the 'traditional ways' (Bennis 1966: 44) of evaluating organizational effectiveness seem very crude, since they overlook several features of the question that are now salient. Indeed traditional writers were overly concerned with performance, measured according to more or less rigid standards and with limited 'output characteristics' (Bennis 1966: 41). Certainly Bennis raises an important problem in contemporary organization theory: the traditional concept of organization effectiveness reflects a very narrow view of the 'determinants' (Bennis 1966: 44) of the organization. However, the assumption that the concept of organizational health brings any clarification or broadening of organization theory is questionable. Besides being foreign to the scientific study of formal organizations (for reasons to be explicated later), the notion of organizational health fails to solve the theoretical and operational problems raised by the old concept of organizational effectiveness, and creates new ones as well.

My main contentions in this regard can be summarized as follows:

1 / Organizational health as conceptualized by Bennis is extraneous to the field of organization theory. It is a mechanical extrapolation of an attribute that may be pertinent to the life of the individual, but not to the nature of the formal organization. Bennis's concept of organizational health implies the reified existence of a collective or organizational mind, whose organismic overtones can hardly be shown to be in tune with the framework of contemporary social science. One wonders if it is not a recurrence of anthropomorphic fallacies like Gustave Le Bon's 'collective soul' and McDougall's 'group mind.'

2 / The assumption of classic writers, from Taylor to Chester I. Barnard, that there is no organization theory without objective standards for evaluating the specific activities of formal organization continues to be valid today. In other words, since the formal organization is essentially defined by a specific type of rationality – instrumental or functional rationality – which is related to optimization of means in order to reach specified goals, criteria for assessing organizational effectiveness are a serious theoretical issue. Today we may reject as crude and inadequate the approach to organizational effectiveness proposed by Taylor and 'scientific managers.' However, in any attempt to tackle this problem their speculations and findings deserve serious consideration. Organizational effectiveness has features that Taylor and other earlier writers overlooked. The organization theorists' and practitioners' task is not to avoid the question, nor to prescribe placebos. It is, rather, to confront effectiveness head-on in all of its present complexity. The concept of organizational health, I submit, begs the issue.

In trying to substantiate the notion of organizational health, Bennis identifies the organization with a *who* ('it needs to know who it is') constituted by several 'selves,' the task of the executive being 'to strive toward congruence' or 'harmony' of these 'selves,' 'insofar as it is possible' (Bennis 1966: 52, 54).[9] Organizational 'selves,' as suggested by Bennis, are a reclassification in behavioral terms of the different structures that Wilfred Brown sees in any complex organization: the *manifest* structure, as expressed by an organizational chart; the *assumed* structure, as given in the phenomenological perceptions of individuals; the *extant* structure, as objectively detected by the organization analyst; and the *requisite* structure, as the optimal one the organization should have within its limited circumstances. Brown's classification is valid and has explanatory force, but by psychologizing such structures and equating them to 'selves,' Bennis misplaces the concept of self and brings confusion to the terms of organizational analysis. One can indeed assess and depict such structures, but one loses accuracy in understanding organizational complexities if one resorts to the category of self as proposed by Bennis. This is exactly the kind of mistaken approach that Katz and Kahn object to when they observe that 'some recent psychological approaches look closely at people and forget their structured interdependence in the organizational context' (Katz and Kahn 1966: 336). Although Katz and Kahn are psychologists, they are very alert to the pitfalls into which the psychologist may fall when indulging in the extrapolations of concepts to organizational problems. Thus Katz and Kahn state: 'The approach taken in dealing with organizational problems has been oversimplified and too global. People have either assumed that organization was like a single individual, or that here was a single problem of motivation for the entire organization with a single answer, or that the organizational structures and processes could be ignored in dealing with the psychology of the individual' (Katz and Kahn 1966: 336).

Psychologists are welcome in the field of organization theory and practice. Indeed we badly need them, but for them to be useful they have to understand the specificities of the organizational phenomenon, which can never be fully comprehended with categories pertinent to individual psychology. A classic demonstration that psychologists can give outstanding contributions to the development of organization theory is *The Social Psychology of Organizations* by Katz and Kahn.

Moreover one can say that all the inaccuracies of the concept of organizational health derive from a common root: the misplacement of the concept of mental health. Indeed if mental health is a valid concept (and some question that it is), its standards are applicable only to individuals, and can never be applied to collectives or deduced from the organizational setting. The concept of organizational health is directly related to the psychology of adjustment, and does not acknowledge individual autonomy. It is not a scientific category but a disguised ideological tool; it is a pseudo-scientific device aimed at total inclusion[10] of the individual within the organization.

When used by practitioners and consultants as a referent for intervening in organizations, the pseudo-concept of organizational health may lead to stifling the individual's psychological energy. In their interventions such practitioners confessedly aim at integrating the individual and organization. This is actually a sinister endeavor which can be carried out only at the expense of the substantive dimension of persons. The kind of psychology underlying the practice of these 'integrationist' consultants is based on a misunderstanding of the nature of socialization and of the organizational phenomenon itself. It is a motivational kind of psychology which implies that the behavioral syndrome[11] inherent in the market-centered society is equivalent to human nature at large. Motivation thus understood becomes tantamount to control and repression of the individual's psychic energy. Nevertheless this type of psychology forms the conceptual framework of some practitioners, educated in our schools of business and public administration, who claim to possess skills for 'managing human stress.' Only the fact that they are victims of a deceptive and naïve teaching may excuse them from the charge of acting as embezzlers of their credulous clients.

A scientific psychology does not necessarily comply with meanings coming from institutionalized definitions of reality. It acknowledges a depth of psychic individual reality which resists being totally captured by social and organizational definitions.[12] Relations between individuals and organizations always imply tension and can never be 'integrated' without crippling psychic costs. Formal organizations are nothing but tools. Individuals are their masters. If psychology is to be a component of the conceptual framework of practitioners and consultants – as it must be – more sophistication is needed in its teaching in our schools of business and public administration. It is encouraging that such a re-orientation is already being suggested in recent writings appearing in scholarly journals.[13]

PERSONS AND SYSTEMS MODELS

Regrettably, without explicitly subscribing to it, the 'integrationist' ideology pervades a great deal of what systems designers and analysts do as consultants and practitioners. Very often they lose sight of the necessary tension between persons and contrived systems by relying on a too holistic conception of system. In general they reify the organizational system, i.e., they emphasize the dependence of the parts of the whole, instead of accurately dealing with the interdependence of the internal and external constituents of the whole. Robert Boguslaw addresses this point in his book *The New Utopians*. Systems designers of mechanistic and organismic orientation take for granted the operative rules inherent in institutionalized systems and 'in the light of the status quo ... proceed to explain how human groups can or do adapt to the world in which they find themselves' (Boguslaw 1965: 3). In practice this is tantamount to allowing the operative rules of formal organizations to fashion the citizens' needs for food, shelter, clothing, transportation, education, and leisure. This bias is aggravated when combined with the misuse of cybernetics which, as Sheldon Wolin has put it, 'consists in likening the nature of human thinking and purposive action to the operation of a communications system, e.g., "the problem of value is like" a "switch-board problem," or " consciousness" is analogous to the process of feedback" (Wolin 1969: 1076). Even a cybernetician aware of the fallacies of mechanistic and organismic analogies like, for instance, Karl Deutsch, is caught up in the practice of misplacement of concepts. Thus he employs a mecanomorphic image in defining the polity as a 'steering system' and statemanship as the 'art of driving an automobile over an icy road' (Deutsch 1966: 182–5). Accordingly the systems analyst is only concerned with polity's capabilities to accomplish its goals; the ethical dimension of such goals is none of his business. The 'threatening' outcomes of such cybernetic posture has been highlighted by Giovanni Sartori (1970: 1035–6). Concerns identical with those of Wolin and Sartori inspire Habermans's discussion of systems analysis (1970: 106–7).[14]

My intent here is not to dismiss systems models, but to argue against their unqualified utilization for administrative analysis and design. In principle systems models have a utility in the administrative domain mainly when the structure-maintenance functions of systems are legitimately to be monitored and nurtured. But when focusing upon structure-elaborating and changing functions of systems, analysts should be prepared to deal with the true nature of systems dynamics, of which the tension between persons and social structures is constitutive.

A significant attempt at refining systems analysis has been undertaken by contemporary writers.[15] For instance, Arthur Koestler is proposing such a refinement when he presents the concept of *holon* as a tool for systems analysis which goes 'beyond atomism and holism' (Koestler 1977, 1969). He sees organisms and organizations not as absolute wholes, but as constituted by sub-assemblies – organelles, organs, organ-systems – each having a remarkable degree of autonomy and self-government. Each of these sub-assemblies is a *holon*, 'which has two faces looking at opposite directions – the face turned towards the lower levels is that of an autonomous whole, the one turned upward that of a dependent part' (Koestler 1969: 197). One can argue, however, that the incorporation of the concept of *holon* into systems analysis does not seem to eliminate the latter's reductionist and holistic bent. One might say that the concept of *holon* implies a functional account of personal consciousness, either as 'parts relative to superstructures' or wholes 'relative to sub-structures' and that this view evidently does not express what a person is traditionally supposed to be. A person is not essentially a functional constituent of a system. I think that Kant's definition of a person ultimately still holds true: 'A person is subject to no laws other than those that he (either alone or at least jointly with others) gives to himself' (Kant 1965: 24). Thus a person may eventually find himself in a system without necessarily being a functional part of it. A person in a contrived social system may very well be a Trojan Horse, i.e., a deliberately disguised agent of disruption both of superstructures and substructures.

Attempts at integrating the individual and the organization are based on a misunderstanding of the nature of a person. I submit that only a delimitative view of organizational design can counter the unqualified practice of systems analysis.

Misplacement of concepts permeates contemporary writings about organizational problems and issues. As a result the citadel of today's organizational scholarship is like a tower of Babel. The confusion of tongues is almost deafening. The source of much of this confusion is the distortive language which has emerged as a consequence of the ascendance of economizing criteria into the social fabric at large and the diffusion of the political into the social. The impact of these developments on language will be considered further in the following chapter. This organizational dilemma cannot be overcome unless administrative theory refuses to assume that the criteria inherent in formal organizations are the criteria of human existence at large. Instead organization theory should become an inquiry into multiple types of social systems, of which the formal economizing setting is a particular case.

CONCLUSION

Paradoxically the disciplinary field of organization theory had a much clearer sense of its assignment before the rise of the so-called human relations school in the thirties. David Riesman and W.H. Whyte should be read again, for they cogently explain how the human relations school was triggered by the imperatives of a business structure demanding emphasis upon consumption rather than upon saving.[16] From Taylor to Luther Gulick professional administrators were very much on the track of what the systematic study of laboring activities and productivity should be. Thanks to them some basic permanent issues of administrative science were identified. They can be articulated as follows:

1 / Labor and productivity are systematic objects of scientific study. Peter Drucker correctly considers Frederick Taylor as a pioneer of today's 'knowledge economy' according to which 'the key to productivity [is] knowledge, not sweat' (Drucker 1969: 71). This can be said of the so-called classical school at large.

2 / There is no science of formal organization without technical procedures to measure and evaluate labor outputs.

3 / Jobs should be technically designed. Their designers should take into account man's physiological and psychological condition. It is not true that Taylor and the classical school neglected the human factor in organizations. What is to be highlighted is that their conception of man was reductionistic and too limited.[17]

4 / Human capabilities are not 'intuitively obvious either to the worker, or to the observer';[18] they must be technically and experimentally detected.

5 / Performance on the job cannot be improved and efficiently organized without systematic training of the workers. In other words, technical training does not necessarily remove or stifle, but rather accentuates human differences.

These are some permanent issues of the theory of formal organization. The classicists' approach to these issues may legitimately be criticized for being theoretically shallow. But at least they realized that formal organizations are not the appropriate settings for people's de-alienation and self-actualization. They were more conscious of their limits than the current 'integrationists' and 'humanists' who claim to know how to design 'authentic,' 'proactive,' and 'healthy' organizations. As harbingers of the 'knowledge industry,' the classicists have undertaken a historical mission, which was to provide the American economy and society at large with sophisticated, capital-intensive, productive capabilities. They handsomely fulfilled this mission. In significant measure their inventiveness and ingenuity have contributed to the United States becoming the first tertiary and service economy in the history of mankind – an economy in which formal organizations, so to speak, can take care of themselves with comparatively little help from people. Today the fundamental mission of organizational theory professionals is not to legitimize the total inclusion of people within the confines of formal economizing organizations, but to delimit the scope of such organizations in human existence at large. The time is ripe for the practice of an unprecedented kind of organizational science which is sensitive to the diverse issues of human life, and which is able to deal with them in settings where they appropriately belong. This, as well as the next chapter, which deals with cognitive politics (a phenomenon related to misplacement of concepts and issues), paves the way for an analytic discussion of the epistemological underpinnings of the new science of organizations.

5

Cognitive Politics: The Psychology of the Market-Centered Society

The so-called science of organization, as we now know it, is entrapped within unchallenged assumptions derived from and reflective of the market-centered society. As long as it remains uncritical of itself, misplacement of concepts and cognitive politics will adversely affect the practice and teaching of the administrative discipline by constricting any effort at true theory building in this domain. I have already explained the notion of misplacement of concepts. In this chapter my purpose is to articulate the nature of cognitive politics.

It should be pointed out that little, if any, attention has been given by organizational theorists to the political dimension of cognition. Politics and cognition have traditionally been treated as separate and distinct areas of study. This situation is reminiscent of an earlier historical period when organizational theorists claimed a clear-cut separation between politics and administration. At that time theoreticians and practitioners lacked the conceptual tools to identify political processes within the organization. This state of administrative scholarship changed when new social circumstances made this deficiency intolerable. Following this stage, when politics came to be recognized as an inherent dimension of activities taking place in organizations, political activity was incorporated into organizational theorizing, but even here politics was understood only as a struggle for power through processes of rewards allocation.

Today, I submit, contemporary developments make the separate and distinct study of cognition and politics inexcusable. This is because the influence of cognitive politics, once restricted to marginal enclaves within the larger social fabric, has today become pervasive. The cognitive patterns required by the concerns of expedience in the market place have become tools of cognitive politics, particularly induced by formal organizational structures and strategies, extending nowadays to society at large. Even to this day a systematic examination of politics as a dimension of cognition has been ignored by organizational theorists. The idea of cognitive politics remains extraneous even to those who long ago abandoned the old dichotomy between administration and politics.

Cognitive politics, to offer a preliminary definition, consists in a conscious or unconscious use of distorted language, the intent of which is to induce people to interpret reality in terms that reward the direct and/or indirect agents of such distortion.

COGNITIVE POLITICS, A HISTORICAL DIGRESSION

Cognitive politics is a perennial historical phenomenon. It is a matter treated by Plato in many of his dialogues on the nature and uses of rhetoric. It is well known that Plato expressed his dislike for rhetoric as practiced by the sophists for the very reason that it aimed at producing only beliefs, not knowledge. Nevertheless Plato attempted to save the phenomenon of rhetoric or political persuasion by calling, in several dialogues, but most specifically in *The Phaedrus*, for a dialectical rhetoric, a rhetoric in the service of philosophy. And in the *Gorgias*, an earlier dialogue, Plato, through Socrates, indicates that the typical rhetorician 'has no need to know about the truth of things, but to discover a technique of persuasion' (459c). This kind of rhetoric, Socrates says in his famous comparison, 'is to justice what cookery is to medicine' (456c), i.e., rhetoric is a technique for flattering the crowd and for use with untrained thinkers, since the sophistic rhetorician cannot be convincing among those who have knowledge. Socrates, at one point in the dialogue, refers to rhetoric as 'the resemblance of a part of politics' (463d), a relationship Aristotle elaborates upon in his *Rhetoric*. He allows, however, for the possibility of a rhetoric that issues in knowledge or one concerned with 'making citizens as good as possible' (513c). It would, therefore, produce informed citizenry and not political dupes and deceivers. Thus, Socrates' own use of rhetoric in the dialogues is in the service of the true political actor, i.e., statecraft. As already indicated Plato also suggests that the way to save rhetoric is to turn it into a part of dialectic as he understood it.

It should be pointed out that Plato does not ultimately betray his condemnation of sophistic rhetoric , because in the *Laws* he suggests that a *civil theology* should accompany legislation protecting a sound political system against agents and agencies of disruption. This creed Plato conceives as a distillation of the minimum norms common to all religions, norms whose validity are made evident through rational debate. A word about the nature of a civil theology may help to clarify the functioning of cognitive politics.

Throughout history civil theologies have been legitimate tools for enhancing the endurance of political systems. With less exacting rational scruples than Plato, for instance, one finds Polybius (1972) praising the Roman statesman for introducing among the people, for the sake of the cohesion of the state, 'notions concerning the gods and beliefs in the terror of hell,' a course of action which would not be necessary if the 'state [were] composed of wise men' (Polybius 1972: VI, 56, 6–11). Today, 'constitutional democracy,' although containing a minimum rational element, is more akin to a variety of civil theology in most Anglo-Saxon countries.[1] Civil theology should not be mistakenly identified with cognitive politics, however. A civil theology is expressly formulated not to deceive people, but rather to legitimize a type of social order in terms and images accessible to the comprehension and educational level of the citizenry. The relevant distinction turns on the notion of rational debate. One might legitimately engage in rational debate in order to validate civil theologies, but the indoctrination or subliminal inculcation of distorted definitions of reality which cognitive politics promotes is never a topic for debate among its victims.

In his dialogues Plato sets about the task of developing an art of rational debate that is further refined and codified by Aristotle in his *Rhetoric*. Aristotle considers rhetoric in relation to other disciplines as an 'offshoot of dialectic and also ethical studies.' He further elaborates upon this relationship when he advises the reader that 'ethical studies may fairly be called political,' and for this reason rhetoric sometimes 'masquerades as political science' (Aristotle 1954: I, 1356a, 2, 25). Aristotle is thus keenly aware of the relation between the power of the word and the many masks worn in the name of political legitimation. The thrust of the *Rhetoric* is consistent with Plato's project of purifying rhetoric of sophistic distortion, if we note that what distinguished the rhetorician from the sophist is the moral purpose in the use of rhetoric by the individual.

The rhetorician is a speaker trained in the uses and occasions for the art of persuasion. Substantive morality is a quality of persons and resides in the speaker. Departures from this grounding result in a kind of conduct which abstracts from the tension implied by substantive reason, and lead to the primacy of expedience over morality. Hence efficacious restraints (which is to say ultimately no restraints at all) are the only reins holding in check the speaker's ability to use his skill to deceive or to induce others to issue faulty judgments or indulge in unethical conduct. The acknowledgment of the ambiguous character of language is at least as ancient as the Greeks. The responsible speaker can strive to use his acquired skill to overcome ambiguities of motive or purpose. Although Aristotle occasionally yielded to the sophistic temptation of compiling linguistic recipes, the overall intent of his *Rhetoric* is to subsume the art of persuasion under ethical standards and to explain its many legitimate political uses.

Plato and Aristotle were not the only learned Greeks aware of the phenomenon of cognitive politics. But in Greece the social scope and impact of cognitive politics could be kept under the control of prevailing mores and folkways, as well as exposed in the informal and formal groups where Greeks learned the duties of citizenship. Philosophy and systematic education also worked as countervailing forces against the proliferation of cognitive politics. In other ancient societies, where philosophy properly did not exist, cognitive politics never became an issue since the individual was evidently guarded against such entrapment by his compact mythic experience of reality.

Within their specific cultural grounds individuals could develop a sense of community life free from the influence of cognitive politics. In the lore of most pre-industrial societies one finds exposed in proverbial terminology the common perception of the market place as the site of cognitive politics and deceptive language. In many archaic and ancient societies the market was assigned a prescribed function within strict geographical boundaries away from the mainstream of social life, lest it undermine the basis of community and distort the nature of communication. This historical arrangement was consciously elevated into a differentiated principle of societal design by classic political thinkers such as Plato, Aristotle, and Saint Thomas, who all agreed that in order to preserve the good character of the polity, the market and its peculiar cognitive and linguistic patterns must never be allowed to expand beyond its circumscribed site.

COGNITIVE POLITICS AND THE MARKET-CENTERED SOCIETY

Cognitive politics is the psychological 'coin of the realm' of the market-centered society. It is not merely incidental that in every society where the market has become a centric agency of social influence, community bonds and specific cultural traits are undermined or even destroyed. In light of the consistent pattern of consequences which has followed the diffusion of the market mentality throughout the contemporary world, it is difficult to understand how this phenomenon has escaped systematic investigation. I am here only calling attention to the obvious, but it is precisely the obvious which it is the business of cognitive politics to obscure. This nearly universal event, resulting from the political expansionism of hegemonic market-centered societies, is legitimized as a basic principle of contemporary social science. Thus, the passing of the traditional society (Lerner 1958), the homogenization of human behavior on a world scale (Alex Inkeles 1960, Deutsch 1953), and the identification of modernization with the diffusion of the institutional and psychological requirements of the market (Parsons 1964, McClelland 1961)[2] are all interpreted with normative overtones by conventional social scientists. This circumstance illustrates the parochialism of such a 'science' and its captivity within the framework of the cognitive patterns inherent in the market-centered society.

One is therefore led to inquire into the reasons why the study of cognition as a thrust of politics – or better, politics as a dimension of cognition – has not been an academic subject. We can easily understand the meaning of expressions such as the 'politics of oil,' the 'politics of transport,' and the 'politics of pollution' without extensive analysis. But the meaning of an expression like 'cognitive politics' does not make itself evident without further clarification. Perhaps one reason the expression may at first sound puzzling is because it is in the very nature of 'cognitive politics' to be something hidden. There is no reason of convenience for those who are involved in the 'politics of oil,' 'transportation,' and so forth to deny the fact. This case, however, does not hold for those consciously or unconsciously involved in cognitive politics which aims at affecting people's minds. Should they recognize the intentionality of their activities, not only would the effectiveness of their doings as cognitive politicians be weakened, but serious ethical questions about their goals would be raised as well.

The phenomenon of cognitive politics requires explicit academic investigation if theory building in the sciences of social systems design and organizational management are to be freed from the constraints imposed by the uncritical internalization of market-centered society assumptions. The study of this phenomenon must draw from existing resources in the sociology of knowledge, linguistics, cognitive psychology, cognitive anthropology, and communication theory. A central issue of the study of

cognitive politics is the set of epistemological rules inherent in the prevailing political framework of advanced industrial societies, rules which are
uncritically internalized by the common citizen through the socialization
process and/or through his/her exposure to systematically contrived
influences.

Agents of cognitive politics vary in the awareness of their roles. The
most conscious of them are currently in the news media and advertising
business. The press, radio, and television are in general engaged in a
continuous process of the deliberate definition of reality. The media are
utilized as weapons in the competition for influencing people's interpretation of reality. Both the setting within which the information is
delivered and its linguistic mode are contrived to deceive rather than to
enlighten the public. One hour of television is enough for anyone to
realize that the politics of cognition is an overwhelming fact of contemporary life. The successful selling of a product is today not so much a
result of the accurate understanding of its true properties by consumers
as the outcome of a disguised political battle against common sense. In
fact the present structure of consumption in this country, where large
numbers of people are induced to believe that they want (and therefore
should buy) what they do not need, is made viable thanks to the practice
of cognitive politics. The process of formal education is also largely
conditioned by this sort of politics. Sometimes individuals and groups
awake to the politics of cognition. Thus, for instance, women, blacks, and
chicanos in this country are no longer willing to accept their images as
they have typically been represented.

Cognitive politics is a fundamental part of formal organizational settings of all categories and sizes. Each formal organization has its specific
jargon which is an important stabilizing and buffering device. Such
jargon contains a certain cluster of tacit rules of cognition or definitions of
reality which are transmitted to the organization's members in the process of socialization.

Moreover, organizations currently typical of the market society are of
necessity phony and deceitful. They are bound to deceive both their
members and clients, inducing them on the micro-level not only to accept
their output as desirable, but also, on the macro-level to believe that they
exist and function in the vital interests of society at large. Organizations
today have an unprecedented and active role in the individual's socialization process. They attempt to become *the* society. And they would seem
to have the ability to do so because they are powerful epistemological

systems in themselves and are presently unrestricted in influencing citizens through the exercise of cognitive politics.

In pre-industrial societies formal organizations had little bearing upon the process of socialization of the individual. To be sure, in those societies the mores and folkways under whose influence the individual learned a particular world view and the standards of right conduct were by and large free of the contrived conditioning of formal artificial systems. The person would learn to become a member of society by participating in a variety of groups which in general did not have the instrumental character of formal organizations as we know them today. Moreover, in no society prior to the industrial one had economizing organizations ever assumed a central and deliberate role in the process of socialization. This circumstance is peculiar to the market-centered society mainly in its late industrial stage.[3]

Today the market tends to become the shaping force of society at large, and the peculiar type of organization which meets its requirements has assumed the character of a paradigm for organizing human existence at large. In such circumstances the market patterns of thinking and language tend to become equivalent to patterns of thinking and language at large; this is the environment of cognitive politics. Established organizational scholarship is uncritical or unaware of these circumstances, and thus is itself a manifestation of the success of cognitive politics.

In the following sections I will attempt to substantiate this argument by examining three unarticulated assumptions of current organizational scholarship. They are: (1) the identification of human nature at large with the behavioral syndrome inherent in the market-centered society, (2) the definition of man as a jobholder, and (3) the identification of human communication with instrumental communication.

A Parochial View of Human Nature

Organizational theorists and practitioners have unconsciously been captured within the domain of cognitive politics by indulging in the formulation of concepts and methods as well as in the implementation of management strategies and designs that accept the immediate organizational world as self-evident.

Such a pre-analytical orientation is particularly noticeable in the early historical stage of the science of organizations. Frederick W. Taylor, the founder of 'scientific management,' accepts the set of psychological requirements of the market system as tantamount to human nature. The procedures he prescribes for motivating people in labor settings rest on the assumption that competition, calculability, zest for gain, and purely economic features as such epitomize the essence of human nature. The fictitious character of this notion of human nature is too self-evidently untrue to deserve a thorough discussion. Historical and anthropological data are now easily available which demonstrate that only in the modern industrial society, thanks to institutional imperatives, has the individual been induced to behave as an economic being. By and large, in pre-industrial societies, economic determinants of human conduct never had the institutional primacy they have assumed in the market-centered society.[4] Moreover, Taylor considered 'scientific management' and its motivation correlates as a referent for the design of not only labor settings, but also the family, schools, and all social life. In other words, he envisioned the global social fabric as an enlarged market domain.

Taylorism is now presented in the literature as an accomplished historical stage of the science of organizations. But the definition of man as an economic being, mitigated and disguised as it often is, continues to determine the course of action of organizational designers and policy makers. The macro-institutional framework of the market-centered society is monitored by policies which are predicated upon that definition of man. Established economic science is still the primary source from which strategic governmental policies are derived. In order to succeed in such a society, according to its rules of reward and punishment and its overall criteria of manpower and resource allocation, the individual has to program himself as an economic being,

Government policies are now becoming inoperative since they are increasingly thwarted by bio-physical constraints of production and resource allocation which standard economics systematically overlooks or claims to account for within its conventional framework as 'exogenous variables.' For instance, phenomena like inflation and unemployment no longer respond to conventional government policies, mainly because the normative economic scenario they imply is mismatched to the concrete circumstances of the world. No wonder that conventional economists react to this situation with bewilderment.[5] Regrettably, although a model of policy formulation sensitive to the biophysical constraints of production and resource allocation has recently reached a high degree of theoretical consistency and pragmatic viability, it is still conventional economics that constitutes the wisdom taught in academic institutions and provides the main guidelines for the shaping of Western societies.

Since the advent of the so-called human relations school in the late

twenties, and even today, an increasing number of organizational theorists and practitioners claim to subscribe to 'humanist' approaches to
organizational design. However, when one closely examines such a
'humanism' one finds it deceptive, since in general its representatives
lack a systematic understanding of the spectrum of contextual requirements the practice of humanism should take into account. In other words
and in general, those so-called humanists indulge in the practice of
misplaced issues, a topic which has been analytically discussed in the
fourth chapter of this book.

It is significant and, I submit, not by accident that the widespread
vogue of 'humanist' approaches to management coincided with the stage
in which this country became an 'organizational society.' Let us pause to
make sense of this expression. It is accurate to say that during the decades
in which 'scientific management' was acclaimed, this country was not yet
an 'organizational society.'[6] People would go to formal organizations to
labor and get their income, but this society provided workers and others –
women, young persons below the work age, as well as a few citizens who
did not want to be employed – with plenty of arenas where they could
undertake existential pursuits free from formal organizational pressures.
So to speak, an imaginary yet existentially real boundary line clearly
separated the arena of formal economizing organizations from other
places where a variety of human endeavors were free to follow their logical
course of development. Significantly, during this period *saving* was
emphatically stressed in American life. This circumstance in itself promp-
ted citizens to engage in self-gratifying activities which implied the exertion of their potential as human beings, not endlessly consuming, and
therefore freely spending commodities. When you wanted to save you
stayed home and engaged in indoor and outdoor activities, and discovered the joy of doing things on your own. Formal organizations were
sober and conscious of their limits, and contained themselves within a
delimited context of the citizens' overall life-space. One might say that
during this period the consumer in this country still enjoyed a large
degree of sovereignty in the market system. There existed in the family a
significant reservoir of 'craft competence' (Leiss 1976) which enabled the
citizen to produce a considerable amount of goods he did not find worth-
while or available to buy. Thus the market had to be somewhat sensitive
to the citizens' substitutive craft capacity in order to plan its line of
production.

The rise of the so-called human relations school in the late twenties and
its fast propagation in the last few decades reflects a transition in the

American economy. At an exponential rate, formal economizing organizations and activities increased and differentiated themselves and thus progressively pre-empted the total arena of the citizens' life-space. The emphasis of the American economy today is no longer on *saving* but on *spending*.[7] Increasingly commanded by formal economizing organizations, the media intrude into the private life-space of citizens and entice them to diversify their wants and to express them in such specific terms that only through buying specific commodities can they be satisfied. Through such a process the citizen was bound to lose his craft competence, his leverage to affect the market's lines of production. The nation became an 'organizational society' and the person an 'organization man.' The producer became sovereign in the market and the main actor in the process of manpower and resource allocation.[8]

An account of this transformation has been undertaken by several analysts. There is striking agreement that the existential entrapment of the American citizen is induced by a too contrived social environment in which almost no arena of his life-space is left for autonomous personal pursuits. To claim, of course, that mainstream organizational scholars ignore or overlook the contributions of these analysts is absurd. The truth, however, is that prevailing organizational scholarship does not have a theoretically refined and systematic understanding of the organizational design implications of the contemporary, all too contrived, social environment. Contemporary theoreticians and practitioners tend in fact to legitimize the expansion of economizing organizations beyond their specific contextual boundaries by practicing a misplaced and mistaken 'humanism.' Through 'integrationist' strategies, i.e., through strategies which aim at the integration of individual and organizational goals, they strive to transform economizing organizations into homelike social systems.[9] Thus they indulge in the practice of cognitive politics, by which issues such as love, self-actualization, basic trust, openness, de-alienation, and authenticity are brought within the confines of the conventional organization where they only incidentally belong. Fundamental issues of intersubjective life are therefore misconceptualized; the practice of dealing with them in the realm of economizing organizations is theoretically indefensible.

Only an uncritical view of organizational goals and human motivation can explain why 'humanist' interventionists feel at ease in their attempts, for example, to minimize alienation within petty food factories, to improve the human culture of depletive and pollutive industrial complexes, and to enhance the effectiveness of corporations specialized in providing the public with redundant goods and services which only serve to undermine the citizens' sense of their genuine, personal needs. They do not explicitly question the overall dehumanizing and deceptive character of the job structure of the market-centered society, which in itself does not allow a consistent practice of true humanism. Their endeavors tend to be 'piece-meal' or 'patchwork,' and thus they 'tend to miss the forest for the trees.' There are indications that not all interventionists are wholly unaware of the psychological syndrome inherent in the organizational society. For instance, in a popular book, significantly entitled *Up the Organization: How to Stop the Corporation from Stifling People and Strangling Profits*, Robert Townsend proposes a 'non-violent guerrilla warfare' management strategy aimed at 'dismantling our organizations where we're serving them, leaving only the parts where they're serving us' (Townsend 1970: XII). Townsend considers the 80 million citizens who have jobs as 'psychiatric cases' (Townsend 1970: 121). He explains: 'We have become a nation of office boys. Monster corporations ... and monster agencies ... have grown like cancer until they take up nearly all of the living working-space. Like clergymen in Anthony Trollope's day, we're but mortals trained to serve immortal institutions ... This is not our natural state' (Townsend 1970: XII).

The fact that Townsend's book became the number 1 bestseller when it was published in 1970 and remained seven months on the *New York Times Best Seller List* indicates that millions of citizens are conscious of their existential entrapment in the present industrial society and are open to alternatives. Their psychological propensity to find alternatives to their situation may signify that the moment is now ripe for the rise of a new paradigm of organizational science. One essential task of this new science is a conceptualization of the variety of basic human pursuits and their corresponding specific settings, of which economizing formal organizations are a case limit. That is, it is essential to release the conception of human nature and related human endeavors from the strictures implied by the behavioral syndrome, and to develop the operational approaches needed to design, implement, and nurture a variety of pursuits according to their peculiar goals.

The Joyful Jobholder, a Pathological Casualty of the Market-Centered Society
A basic dimension of contemporary organizational scholarship, illustrating its character as an instance of cognitive politics, is its unqualified assumption that formal job settings are appropriate for human actualization. This assumption is clearly unwarranted for anyone abreast of the

authoritative literature about the nature of the market society. One can hardly justify that a responsible organizational scholar should ignore what Max Weber wrote about the historical peculiarities of the market society and their bearing upon the job structure. Indeed, Max Weber stressed that compared with societies which it succeeded, the market society is a peculiar historical configuration precisely because it cannot function effectively unless the individual's performance as a member of labor settings is impersonal. Social systems congenial with personal performance in labor settings as well as with personalized treatment of their clients hinder the advent and development of the market system. Weber intimated, for instance, that one of the reasons why the Germany of his time was lagging behind Great Britain and other Western European nations was that its administrative framework was still arrested within the patrimonialistic type of labor performance characteristic of feudal social systems. To modernize Germany, he meant, would be tantamount to speeding up the formation of a German national market, and therefore to dismantling the feudal type of administrative framework. Weber's statement has been subject to a great number of qualifications, but the essence of his analysis of modernization as synonymous with the development of the market society still holds true without qualification.

In a market society the efficient jobholder must be a depersonalized actor. He is supposed to abide by the superimposed prescriptions which define his role. One feature of his normal pathology is what Dewey called 'occupational psychosis,' which results from an uncritical compliance with the prescriptions of his occupational role (Merton 1967: 198). As noted by Robert Merton, the jobholder is bound to conform to a 'stereotyped behavior' which 'is not adapted to the exigencies of individual problems' (Merton 1967: 202).

Let us pause here and listen to an autobiographical statement of one corporate actor. His book, *Life in the Crystal Palace*, contrasts the corporate man's performance on the job with the role played by an artist on the stage. While the good actor *projects* himself into his role, the efficient corporate man *hides* himself in identical situations. This contrast between revealing and hiding is instructive: 'The actor must incarnate the role in terms of his own personality. He builds himself into the character rather than merely "playing" the part. In contrast, procedures and protocol at the [corporation] teach us to play our parts with inner dishonesty. Just as the actor pours himself into his role, the [corporation] executive pours himself *out*' (Harrington 1959: 144).

The individual's deeds as a jobholder are incidental to his genuine personal actualization. If a person allows the organization to become the primal referent for existence, he loses contact with his real self and instead adapts himself to a contrived reality. Contrived systems like formal organizations have goals which, only by accident and secondarily, bear upon a person's actualization. True actualizers are the actors able to maneuver in the organizationally contrived world, serving its objectives with mental reservations and qualifications, all the while leaving some room for the fulfilment of their unique project of existence. There is therefore a continuous tension between contrived organizational systems and actualizers. To claim that the individual should strive toward the elimination of such tension, thus arriving at a homeostatic equilibrium between himself and the organization (an instance of cognitive politics which a motivational psychology postulates on allegedly scientific grounds) is to advise the deformation of the self. Only a defective self can find in contrived systems the adequate milieu for his actualization.

Current integrationist organizational 'humanism' relies on a sociomorphic conception of human actualization. Thus it resists acknowledging that the human psyche contains any substantive element that is not internalized through the process of socialization. This mind-set of the integrationist 'humanist' is defenseless against the eventual pathology of social systems themselves. A reversed mind-set, i.e., a substantive person-centered perspective, is necessary for clinically focusing upon social systems. From the standpoint of such a person-centered psychology, 'man is sick – not just neurotic and psychotic people, but so-called "normal" man too – because he hides his real self in the transaction with others [and] equates his roles in the social systems with his identity and tries to deny the existence of all real self which is irrelevant to role' playing (Jourard 1964: 60–1).

The sociomorphic concept of the human psyche strips the individual of his will to meaning. Indeed, meaning is bestowed upon an individual's existence when it is primarily derived from the prospection of his personal potentialities. This is not to say, however, that in actualizing himself the individual is supposed to give free reign to his psychological compulsions and indulge in an indiscriminate fulfilment of his potentialities. In fact, he has to work against many of them if he is committed to accomplish his lifetime, unique personal endeavor. Self-actualization moves the individual toward inner tension, toward resisting complete socialization of his psyche. One needs to qualify carefully the notion of self-actualization lest it justify the individual in failing to live up to the tension inherent in his existence.

The clash between the individual and contrived social system is permanent and inevitable. This clash can only be eliminated by the death of the human self or its paralysis through overadaptation of social externalities. Furthermore the individual's self-actualization is very often than not an unintended consequence of innumerable courses of action. Paradoxically it is an after-the-fact verification rather than a guaranteed agenda. The more the individual is concerned explicitly with self-actualization, the more trapped he finds himself in the puzzle of existential frustration.

Psychologists such as Jung, Laing, Progoff, and others tell us of a deep domain of the human self untouched by the process of socialization. Those who dare to journey through such a domain – creative individuals, poets, musicians, novelists, artists of many kinds, and even mad people – have left accounts of it. The individual that sociomorphic motivational psychology focuses upon is the one for whom the social world is his only 'center of experience.' He has ego but he has lost consciousness of his self, where unspeakable realities lie dormant (Laing 1968: 133, 158).

The sociomorphic motivational psychology utilized and in turn perpetuated by extant organizational scholarship is itself a feature of the behavioral syndrome inherent in the market-centered society. It assesses the 'normality' and worth of the individual according to the function he performs as a jobholder. It is a kind of psychology which does not transcend a peculiar historical episode, since only in the market-centered society have jobs for the first time become the dominant, if not almost the exclusive category for acknowledging the worth of human pursuits. In such a society, not to hold a job means not to be worthy and even not to exist.

Nevertheless the notion of job, as we know it, is so recent that Webster's dictionary still considers it a colloquialism.[10] In Middle English, *jobbe*, the root of the contemporary word *job*, meant a piece or a lump, and had nothing in common with the holding of any specified formal organizational position. Before the advent of the market-centered society the job had never been the cardinal criterion for defining the social significance of the individual. In pre-industrial societies people were productive and had occupations without necessarily being jobholders. Unemployment as a feature of joblessness was inconceivable[11] in the structural design of these societies since they provided a productive function for anyone they recognized as a member. In these societies, what would constitute a semblance of today's mass unemployment was rather the sporadic result of upsetting events such as droughts, wars, feuds, or plagues. Membership in these societies in itself granted to the individual the possibility of being safe against starvation. Starvation would happen only as a collective phenomenon determined by a natural or social catastrophe that would affect all members of the society.

As Adam Smith acknowledges, the market society necessarily transforms the individual into a jobholder: 'Where the division of labor has been once established,' he says, 'every man lives by exchanging, or becomes in some measure a merchant, and the society itself grows to be what is properly a commercial society' (Smith 1965: 22). In the society he conceptualizes as commercial, the individual can only provide himself with the goods and services he needs through the exercise of a job. For performing a job he is paid a salary, a certain amount of money with which he buys what he can afford. In this society he is left without an occupation in the ancient and primitive sense. At the very moment Adam Smith was writing the *Wealth of Nations* he could still see in England areas untouched by the market system. But he laments the fact that 'when the market is small, no person can have any encouragement to dedicate himself to one employment' (Smith 1965: 17), as for instance 'the porter' in 'a great town,' where the market is ubiquitous. He writes:

In the lone houses and very small villages which are scattered about in so desert a country as the Highlands of Scotland, every farmer must be butcher, baker, brewer for his own family. In such situations we can scarcely expect to find even a smith, a carpenter, or a mason, within less than twenty miles of another of the same trade. The scattered families that live at eight or ten miles distance from the nearest of them must learn to perform themselves a great number of little pieces of work, for which, in more populous countries, they would call in the assistance of those workmen. Country workmen are almost everywhere obliged to apply themselves to all the different branches of industry that have so much affinity to one another as to be employed about the same sort of materials. A country carpenter deals in every sort of work that is made of wood: a country smith in every sort of work that is made of iron. The former is not only a carpenter, but a joiner, a cabinet-maker, and even a carver in wood, as well as a wheelwright, a plough-wright, a cart and waggon maker. The employments of the latter are still more various. It is impossible there should be such a trade as even that of a nailer in the remote and inland parts of the Highlands of Scotland. Such a workman at the rate of a thousand nails a day, and three hundred working days in the year, will make three hundred thousand nails in the year. But in such a situation it would be impossible to dispose of one thousand, that is, of one day's work in the year. (Smith 1965: 17–18)

Thus classical economics has been conceived by its founders as a discipline which envisions the job as the cardinal criterion for manpower and resource allocation. Smith saw those areas of England untouched by the market as a drawback to civilization. 'Civilization' would be served by allowing the market to expand in England and eliminate any room for the permanence of characters such as the 'country carpenter' and 'country smith' he describes with pejorative overtones. There should be no reason to worry about the impingement of the expanding market upon the lives of the 'country carpenter' and the 'country smith' who were not trained to perform as jobholders. In the long run they would learn the skills needed to make them part of the emerging type of labor force, and the law of supply and demand would provide jobs for all individuals willing to labor. Smith and the majority of classical economists did not conceive of involuntary unemployment. And, in general, such an assumption was supported by the facts during the approximately first one hundred and fifty years of the market society.

In today's Western economic system, however, there are realities which Smith and the classical economists could not anticipate. In order to get an insight into the nature of these realities, it is helpful to distinguish between primal and demonstrative goods and services.[12] The first are those which meet the limited biophysical needs of food, shelter, clothes, transportation, and elementary services to help the individual maintain himself as a healthy organism and a functional member of society. Demonstrative goods and services are those primarily meant to meet the individuals' desires to express their relative rank in the status structure, their desires being conceived socially and as unlimited.

In retrospect, Smith's vision of the development of the market society as a civilizing process can be justified. As long as the market remained bounded (as in all pre-industrial societies), the provision of primal goods and services was the essential goal of the productive system. In such societies the productivity of the labor force was low and only a dominant minority was able to engage in activities of a civilizational nature. The development of the market system would ultimately bring about abundance, and therefore a more equitable social structure by largely freeing the labor force from the burden of drudgery. Such a *post hoc* justification of the development of the market society was systematically articulated by John Stuart Mill. 'The legitimate effect' of the market society, he intimated, is the 'abridging of labor.' At some point in the development of the market society any 'progress in wealth' would be a 'postponement' of a 'better distribution' of goods and services. Accordingly he states: 'I confess I am not charmed with the ideal of life held out by those who think that the normal state of human beings is that of struggling to get on; that the trampling, crushing, elbowing, and treading on each other's heel which form the existing type of social life, are the most desirable lot of humankind, or anything but the disagreeable symptoms of one of the phases of industrial progress.'[13]

This retrospective explanation of the development of the market society sheds new light on interpreting events characteristic of contemporary economic history. One might argue, for instance, that the Great Depression in the United States was an indication that in this country the market system had accomplished its historical role. Paradoxically the Depression did not mean lack of capability to produce plenty of primal goods for all citizens, but resulted from people's low acquisitive power to buy them. Just when the Depression was about to become visible, President Hoover said: 'We shall soon with the help of God be within sight of the day when poverty will be banished from the nation.'[14] In other words, this country's market had developed a capital stock and a technological logistics capable of producing an amount of primal goods and services necessary to meet the basic needs of the population. Moreover, it could realize this accomplishment without requiring that each individual be a jobholder. But the prevailing mind-set of private entrepreneurs and public economic policy makers precluded them from understanding the concept of job as anything other than a mechanism for distributing income. This in turn led them to believe that there was no way for such a primal economy to provide occupations for all.

Classical economists were conceptionally unprepared to understand and overcome this crisis. The Keynesian revolution consisted in rescuing the discipline of economics and overcoming the market impasse by tapping, through spending, the potential of the market to produce demonstrative goods and services. Under this new condition the market again could be able to provide jobs for all, which in turn would increase purchasing power. One should notice, however, that Keynes conceived of 'job' as the cardinal criterion of manpower allocation. His mind was captive within the structured social design implied in this organizational principle. For this reason he was still a classical economist and failed to produce a true 'general theory of employment,' if we understand by employment a condition in which the individual can exercise a socially useful production without necessarily being a jobholder. Granted, Keynesian economic policies did rescue the market and did restimulate its activities. But Keynesianism was only a temporary postponement of the crisis, foreshadowing the closure of the historical validity of the category 'job' as an organizational principle of production.

Today in market societies, despite the fact that the production of demonstrative goods and services parallels if not exceeds the production of primals, the market is again becoming unable to provide jobs for those willing to labor. This has become a secular structural trend which defies any system of economic policies including those Keynesian in nature. Thus, for example, in *Work in America* one reads:

The economy itself has not been [able] ... to absorb the spectacular increase in the educational level of the workforce. The expansion of personnel, technical, and clerical jobs absorbed only 15% of the new educated workers; the remaining 85% accepted jobs previously performed by individuals with fewer credentials ...

If matters continue as they are, the disparity between the supply and demand of educated workers is likely to be exacerbated in the next decade. About ten million college graduates are forecasted to enter the job market in the period, while only four million graduates will leave the workforce through retirement or death. This means that there will be 2 college graduates competing for every 'choice' job, not to mention the additional 350,000 Ph.D's who will be looking for work. (O'Toole et al. 1973: 135, 136)

To say that the psychologically dysfunctional character of the job structure dominant in advanced industrial situations has been unnoticed by organizational scholars is not completely accurate. But the fundamental theory of such dysfunctionality and its impingement upon the life of the citizenry at large has remained sadly neglected. To be sure, an urge for personalization is now a dominant characteristic of the psychological profile of citizens. Reporting the findings of a national survey undertaken in the early years of the past decade, the authors of *Work in America* state: 'Compared to previous generations, the young person of today wants to measure his improvement against a standard he sets for himself. [Clearly, there is much more inner direction than David Riesman would have predicted two decades ago.] The problem with the way work is organized today is that it will not allow the worker to realize his own goals' (O'Toole et al. 1973: 51).

Extant organizational scholarship is apparently also sensitive to this trend, but mistakenly focuses its attention on meeting the citizens' need for personalization within job settings. This implies a double misunderstanding. First, conventional or organizational theorists and practitioners do not realize that jobs themselves are incidental to the process of personalization; second, apparently they do not take account of the fact that the job structure of advanced market society is chronically unable to provide occupations for all citizens willing to labor.

Complaining that the business structure now offers to individuals tasks which are not designed to meet their needs of personalization is less than one should expect of a true humanist approach to social systems design. The crux of the matter is that jobs are no longer the only means to engage individuals in socially meaningful productive activities. The fact that in the market society the holding of a job is the only avenue for the individual to become a socially significant person has to be interpreted as a functional requirement for developing the logistics capable of producing plenty for all. As the handmaid of such a logistics, the market system would not have fulfilled its historical assignment without the concept of 'job' becoming the essential principle of the social organization of production. Let us pause to identify the nature of this historical assignment.

The market system in its Western form proved that production of primal goods for all is possible without requiring the involvement of most of the population in laboring activities. The rate of productivity of the laboring activities in pre-industrial systems was so low that leisurely pursuits could only be a privilege for the few. As long as production in these societies was unthinkable as a systematic object of applied knowledge, laboring had to be assigned to the majority of people and justified as a matter of principle and as a fact of nature. Nevertheless in these societies the laborer was free from certain vexatious imperatives which bear peculiarly upon today's jobholder. For example, the pre-industrial laborer was deprived of refined leisurely activities, but he was master of himself as a laborer, not a factor of production to be treated as a commodity priced according to the law of supply and demand. His laboring activity left him ample room for pursuits through which he could freely actualize his personal potential. Even in the Stone Age, as Sahlins (1972) portrays it, affluence was available to the individual.

The ultimate assignment of the market system was to transform production into a scientific activity and to provide society with processing capabilities for a high rate of productivity, simultaneously liberating men from laboring. In the process of accomplishing such an assignment the market society had to use man as a depersonalized factor of production. The deformation of the human self imposed by this transition has been the psychological price paid for the creation of the logistics of plenty of primal goods for all. This is the 'great transformation' to be credited to the market system. More than anything else, a single historical incident proves that such a 'great transformation' has been accomplished. This incident begins with World War II and unfolds through the following thirty years.

The argument is developed by H.F. William Perk. World War II activated an unprecedented productive capacity in the American industrial system. Regrettably an immense quantity of destructive materials and articles were produced and delivered to the 'consumer' on the war fronts. This effort involved a number of costly operations like packaging, transportation, distribution, and their complex administrative correlates. Subsequently the cold war period of the late forties to the late fifties ushered in the age of 'overkill' in which our destructive capacity increased a million times. During this period we also saw the emergence of a non-military industrial system capable of providing material abundance. Some indicators of this highly productive system are the practice of 'planned obsolescence' of consumer articles, the deliberate escalation of the demand for redundant and demonstrative goods and services involving the engagement of a considerable part of the labor force, the acknowledgment that productive capacity could not be used 100% lest surpluses upset the conventional terms of the market, and the underutilization of the combined industrial capacity of Europe, the Soviet Union, and Japan for producing plenty of primal goods and services.

Thus the production of plenty of primal goods could have been undertaken by the then existing logistic system. But this technical possibility did not constitute in itself the requisite and necessary condition for the 'great transformation.' Perk states: 'There is reason to believe that the Great Depression that preceded World War II was as bad as it was because the technical possibility of abundance was already imminent, and the controllers of the industrial system did not know how to cope with this condition' (Perk 1966: 362).

Perk's argument suggests a revaluation of the economic policies, Keynesian in general, which were formulated and implemented to overcome the Great Depression and that constitute even today a great part of the conventional wisdom of standard economics. Apparently the formulators of these economic policies failed to understand that in a stage where abundance of primal goods and services can be produced with a low rate of involvement of individuals in the formal job structure, a reinterpretation of the historical role of the market is needed – i.e., one implying its delimitation, through new political regulations, as an enclave assigned with economizing activities *par excellence*. Instead, policies of Keynesian and similar persuasion postponed the delimitation of the market by expanding its lines of production and activities, thus leading it to pre-empt the management of the social fabric itself.

Cognitive politics is an inevitable dimension of such a pre-emption, and administrative scholarship, by taking the present job structure as a permanent feature of the economy, fails to understand the organizational plight of American citizens.

The Psychology of Instrumental Communication
Standard administrative scholarship fails to see that within economizing organizations, communication is essentially instrumental in the sense that it is systematically designed to maximize productive capabilities. Within such organizations the individual himself is a resource to be efficiently used. Psychology becomes a technology of persuasion for increasing productivity. To blame economizing organizations for being unable to meet the needs of the individual as a unique person is as futile as to blame the lion for being a carnivor. They cannot do otherwise, and since without economizing organizations society would not function properly, they must be realistically understood for what they are. Substantive communication, i.e., communication intended to unveil the subjectivity of persons involved in self-gratifying exchanges, is ill-suited to economizing organizations. Accordingly, to assume that self-actualization can be fostered within economizing settings, as organizational 'humanists' do, is to indulge in cognitive politics. Indeed such an assumption leads to the practice of deceptive personnel development techniques meant to facilitate the total disclosure of people's subjectivity out of context, i.e., in performing roles of an instrumental nature.

In the domain of organizational scholarship one of the rare attempts to come to grips with the concept of communication was undertaken by Herbert Simon in his book *Administrative Behavior*, published in 1947. Simon states: 'Communication may be formally defined as any process whereby decisional premises are transmitted from one member of an organization to another' (1965: 154).

Considering that Simon is essentially concerned with economizing organizations, this statement is indeed accurate. Within economizing organizations communication between people takes place independently of what they are as persons, and elicits from them information which is intelligible only under superimposed decisional premises. In other words, such a species of communication is not free from external imperatives, and does not serve as a vehicle for the individual's self-gratifying personal, subjective disclosure. Simon clarifies his definition as follows:

'Through his subjection to organizationally determined goals, and through the gradual absorption of those goals into his own attitudes, the participant in organization acquires an "organization personality" rather distinct from his personality as an individual. The organization assigns him a role: it specifies the particular values, facts, alternatives upon which his decisions in the organization are to be made' (198).

It is clear why Simon correctly denies the possibility of self-actualization within formal organizational settings.[15] But although his picture of organizational realities is more realistic than the one portrayed by his 'humanist' opponents, it is nevertheless prone to justify the practice of cognitive politics. To some extent Simon's book reflects the social environment of this country in the aftermath of World War II, when an optimistic and uncritical view of the functions of the organization was rampant. Thus in focusing upon the relationships between the individual and the organization, his statement is lopsided in favour of the organization. For instance, he admits that 'the employee signs a blank check in entering the organization' (116). Granted, he cautions that 'the area within which organization authority will be accepted is not unlimited' (117). But his understanding of these limits is wanting. Relying on Barnard he sees the jobholder as a split personality carrying simultaneously an 'organization personality' and a 'private personality' (204). 'When,' he says, 'the organizational demands fall outside [their confines], personal motives reassert themselves, and the organization to that extent, ceases to exist' (204).

Yet Barnard's and Simon's conception of the dual personality of the jobholder is insufficiently qualified. It is presented in overly mechanistic terms and implies a loyalty to the organization which leads to – let us be frank – hideous existential outcomes. For instance, Barnard tells the story of a telephone operator so concerned with her sick mother that she accepted, against her inclinations, a job in a lonely place because from there she could see the house in which her mother lay bedridden. Nevertheless when a day came in which a fire broke out in the house, she watched the incident without leaving her switchboard desk. Barnard praises the operator: 'She showed extraordinary 'moral courage,' we would say, in conforming to a code of her organization – the *moral* necessity of uninterrupted service' (Barnard 1948: 269). It is precisely this type of unqualified loyalty of the jobholder to organizations which eventually transforms them into agencies of moral corruption, inducing individuals, for instance, to accept the Nazist horrors as normal facts of state life, or to indulge in violations of the law like those in which President Nixon and his staff were caught during the Watergate case.

Moreover it is evident that the passive subjection of the individual as a jobholder to the organization has a deep effect upon his personality which does not cease in his private life space. If, as Simon has it, the jobholder is expected 'to relax his critical faculties' in order to 'permit the communicated decisions' to 'guide his own choice' (151), this disposition may doom him to turn his 'occupational psychosis' into his second nature, as some analysts suggest (Merton 1967 and Mannheim 1940). In other words, he will be weakening his capacity to exercise, outside the organization, ethical and critical judgments of a personal nature. The unqualified legitimization of such a bearing of the organization upon the individual is to be acknowledged as an instance of cognitive politics.

Indeed it is questionable if in his off-job time the individual is provided with sufficient areas free from penetration of organized social pressures. The social environment of this nation is highly contrived, if one takes account of how the delivery of information to citizens normally takes place. It has been correctly asserted by many authorities that being highly controlled by gigantic business concerns the mass media in this country largely promote an 'unthinking allegiance' (Lazarsfeld and Merton 1974: 567) to the status quo. Merton and Lazarsfeld interpret this institutional mode of delivery of information as serving 'in place of intimidation and coercion' (556). Caught within such a mode the social environment at large itself becomes a mechanomorphic environment. And by internalizing its norms and requisites the individual is induced to become a mechanomorphic system himself. He substitutes a contrived jargon for common sense and inevitably loses the verbal skills to speak about deep levels of his psyche which resist expression through mechanomorphic signs.

Herbert Simon would see nothing remarkable about the ongoing passive subjection of the individual to the social environment. Contending that there is only a difference of degree between an ant and man he would argue that the individual's 'inner environment may be largely irrelevant to [his] behavior in relation to the outer environment' (Simon 1969: 25). Indeed he states:

An ant, viewed as a behaving system, is quite simple. The apparent complexity of its behavior over time is largely a reflection of the complexity of the environment in which it finds itself ...

I should like to explore this hypothesis, but with the word 'man' substituting for 'ant' ...

A man, viewed as a behaving system, is quite simple. The apparent complexity of his behavior over time is largely a reflexion of the complexity of the environment in which he finds himself.

I myself believe that the hypothesis holds even for the whole man ... A thinking human being is an adaptive system; his goals are defined by the interface between his inner and outer environments. To the extent that he is effectively adaptive, his behavior will reflect characteristics largely of the outer environment (in the light of his goals) and will reveal only a few limiting properties of his inner environment. (24–6)

This statement is highly representative of the conception of man which pervades behavioral psychology. This conception is formulated in two unarticulated steps. First, it defines man as a behaving system. Second, it assumes that a behaving system is equivalent to an information processing system. Simon's 'humanist' opponents in the domain of standard organizational scholarship do acknowledge, in the individual, a gamut of needs grounded in his personal subjectivity. Paradoxically, however, in practice there is no essential polemic between Simon and those who agree with him, on one side, and their opponent colleagues, on the other, for the simple reason that the latter fail to understand that those needs cannot be met within mechanomorphic environments, where they feel at ease to exercise their clinical expertise. 'Humanists,' for instance, believe that trust, authenticity, love, and openness can be fostered in the interpersonal culture of economizing organizations if their members are engaged in feedback sessions where they are encouraged not only to produce information about their feelings, but also to process information about themselves coming from others. This type of group dynamics which includes, for instance, techniques like T-group and sensitivity training, is mechanomorphic and, as sponsored by economizing organizations, unsuited for dealing with issues of personal growth.

Indeed personal growth and personal solitude are inseparable. Personal growth unfolds from within the individual's psyche and most likely is hindered by social or group feedback processes. All socialization is alienation.[16] It is significant that 'humanists' have no hesitation about embracing behavioral psychology. It is unnecessary to restate my critique of this type of psychology presented in chapter 3 of this book. Suffice it to say that behavior is essentially a category of the individual's external life. Naturally a psychology which envisions man as a behaving animal tends to be group centered, or to hinge upon social feedback processes. But this is a very partial view of man's psychic life, of which behavior is a superficial dimension. Essentially man does not behave. As a carrier of reason he essentially *acts*. But behavioral psychology is bound to overlook action as a category of man's inner life, because it is a psychology without reason, i.e., a psychology in which reason is misinterpreted as synonymous with sheer reckoning of consequences. This misunderstanding, predicated upon shoddy scholarship, explains why 'humanists' claim a clash between 'self-actualizing man' and 'rational man.'

The project of humanizing formal organizations is also rooted in that misunderstanding. Indeed, accepting the constricted reason of formal organizations as reason in general, 'humanists' strive to mitigate their intrinsic concern with functional requirements of effectiveness by proposing strategies intended to integrate organizational and individual goals, and thus the contrived and the emotional and spontaneous. Evidently they fail to realize that when engaged in instrumental communications systems, the individual is bound to reject systematically his direct experience of reality. This observation has been lucidly articulated by Joseph Weizenbaum in his book *The Computer and Human Reason*. Weizenbaum correctly points out that the contemporary rejection of direct experience is a skill the individual learns through his socialization, a 'new sense' for him to 'find his way' in a world contrived according to functional requirements of effectiveness (Weizenbaum 1976: 25). In such a world the individual learns to repress spontaneously feelings and insights which would distract his behavior from its instrumental purposes. Thus he eats when the clock says so, not when he is hungry, and he satisfies his other needs in identical fashion. In the domain of organizational scholarship, Barnard's praise of the telephone operator who sticks to her switchboard desk instead of trying to help her sick mother inside a house on fire is a striking illustration of the rejection of direct experience of reality.

The 'humanist' attempt at integrating individual and organizational goals can only be undertaken on the basis of a behavioral psychology (which is hardly more than a cryptic form of cognitive politics) predicated upon a pre-analytic understanding of organizational operative realities wherein the critical functions of standard administrative scholarship are systematically laid aside. The vogue of systems theory is a case in point. As Sheldon Wolin points out, in the systems approach of a behavioral orientation, under the category of 'input,' one reduces heterogeneous issues to a common homogeneous item. For instance, the term 'input' stands 'equally for a civil rights protest, a deputation from the National Rifle Association and a strike by the u.a.w.' (Wolin 1969: 1078).

In the same vein Wolin suggests that such a mechanomorphic orientation, which he calls 'methodism,' has ethical and theoretical underpinnings that eventually lead to 'grotesque educational results' (1078). It is 'ultimately a proposal for shaping the mind' (1064) and its 'assumptions are such as to reinforce an uncritical view of existing political structures and all that they imply' (1064). Very often it teaches 'false rationality and pseudo-excellence,' as another analyst has pointed out.[17] When trained under the influence of approaches of this sort, students in schools of business and public administration are encouraged to comply with a pre-reflective view of organizational realities and thus to become not scholars, but academic clerks.

CONCLUSION

No society in the past has ever been in the situation of today's advanced market-centered society in which the process of socialization is largely subsumed under a cognitive politics exercised by uncontrolled major business concerns. In any society of the past business has never been the centric logic of community life. Only in contemporary modern societies does the market play the role of a centric force shaping the mind of the citizenry. The reform of the market mentality is not a small task. Not only has the citizenry at large become adjusted to it, but also few channels are available through which countervailing influences, intended to undo such an addiction, can be systematically exercised. Captive of a mass communications system managed by major business concerns, individuals tend to lose their capacity to engage in rational debate. By yielding to contrived influences the majority of individuals lose the capacity of distinguishing between the contrived and the real, and instead learn to repress substantive standards of rationality, beauty, and morality inherent in common sense.

I have portrayed standard organizational scholarship as an instance of obliterated common sense. As such, it hardly deserves to be called scholarship at all. Yet it constitutes the core of what is taught in schools of business and public administration. It is also the creed of a professionalism of sorts. There is nothing wrong with professionalism in general. What is regrettable is that it protects itself against rational debate through the imposition to nonconformist thinkers of an opinionated professional consensus. Compliance with this consensus is a condition for their academic recognition.

Contemporary human problems can only be perpetuated, not solved, by cognitive politics. The frontiers of formal organizational theory must be made clear. Rather than placing the formal economizing organization at the center of human existence, one needs instead to focus upon the question of organizational delimitation, of learning to facilitate multiple types of micro social systems within the overall social fabric, turning the formal economizing organization into a restricted and incidental enclave in the human life space, and thus leaving room for interpersonal relationships free from contrived organized pressures. The delineation of a substantive approach to organizations, an approach which is foundational for organizational delimitation, constitutes the thrust of the next chapter.

6

A Substantive Approach to Organizations

Contemporary organizational scholarship has not developed the analytical skills necessary for a critique of its theoretical underpinnings. Rather, it largely borrows skills from without. For this reason it has condemned itself to be pre-analytical and forever on the periphery of social science. A disciplinary field can hardly attain the sophisticated level of scholarship which is worthy of graduate education if it is not capable of critically developing from within itself its epistemological foundations. Focusing upon such foundations this chapter will attempt to present a systematic approach to organizations based upon the concept of substantive rationality.

Formulating a substantive approach to organizations involves two distinct tasks: (1) the development of a type of analysis able to detect the epistemological ingredients of various organizational settings, and (2) the development of a type of organizational analysis expurgated of distorted patterns of language and conceptualization.

Although this chapter deals mainly with the second task, some tentative thoughts about the first are in order.

TASK 1: ORGANIZATIONS AS EPISTEMOLOGICAL SYSTEMS

It is commonly asserted by social scientists that definitions of reality are learned by individuals in the socialization process. As Karl Mannheim points out, when new situations emerge in a society its members normally tend to read them according to established categories. It is as if they 'refuse to admit their novelty' or prefer 'to ignore their uniqueness' (Mannheim 1940: 302). At the level of the micro-organization, March and Simon (1958: 165) call this reactive pattern 'uncertainty absorption.'

When exposed to an unprecedented situation, the individual, lest he jeopardize his psychological security, normally tries to interpret it according to the organization's existing conceptual vocabulary. 'Hence,' March and Simon write, 'the world tends to be perceived by the organization members in terms of the particular concepts that are reflected in the organization's vocabulary. The particular categories and schemes of classification it employs are reified, and become, for members of the organization, attributes of the world rather than mere conventions' (165).

In his commentary about 'uncertainty absorption,' Charles Perrow states that organizations control the action of their members by developing 'vocabularies which screen out some parts of reality and magnify other parts' (Perrow 1972: 152). Given the fact that today's organizations have a 'protean ability to shape society' (199), Perrow asks for a re-examination of the notion of 'environment' as it is currently presented in the literature. Instead of the 'environment' affecting the organization, it seems that the opposite is nearer the truth. The organization must be seen today 'as defining, creating, and shaping its environment' (199). A similar view of the environment is held by J.K. Galbraith (1973) and B. Gross (1973) – a view that they hold to be characteristic of the u.s. social system at large.

Although statements like these are frequent, a systematic examination of their implications has only recently been attempted by a few authors who are concerned with the epistemological dimension of social systems.

Robert Boguslaw confronts this question in *The New Utopians*. He claims that systems design is not a purely technical issue, but should involve a systematic concern for consequences evaluated from the standpoint of human values. However, contemporary system designers focus upon organizational problems using conceptual and operational tools only consistent with the technological status quo (Boguslaw 1965: 4). They work with computer hardware, systems procedures, functional analyses, and heuristics which specify human behaviors and courses of action. Boguslaw tries to lay bare the rules of cognition which dominate the art and theory of conventional systems design, which he finds under the sway of political conveniences. He suggests that systems designers largely rely on 'an after-the-physical fact kind of theory' (p. 2). Accordingly, Boguslaw questions the validity of the 'methods, techniques, and intellectual underpinnings of the various approaches to systems design' (pp. 2–3).

Some systems and communication scholars are also aware of the epistemological questions implied by organization theory. For example,

C.W. Churchman (1971 and 1979) and W. Buckley (1972) have addressed themselves in a high level of abstraction to epistemology as a topic of systems analysis. Similarly, the influence of the computer upon the individual's self-perception has been thoroughly analysed by Joseph Weizenbaum (1976). Yet, until recently specialists in systems theory had not developed conceptual and operational tools to deal with the epistemological system which, although ordinarily hidden, constitutes a fundamental component of any type of organization.

One exception to this trend is Donald Schon. In his view of social system as presented in *Beyond the Stable State* (1971), the epistemological dimension is a systematic topic of concern. According to Schon, any social system consists basically of a structure, a technology, and a theory. The structure is the 'set of roles and relations among individual members' (33). The technology is the actual set of procedures and consolidated practices by which things get done and outputs are produced. The theory is the set of epistemological rules by which the internal and external reality is interpreted and practically treated. In any system these dimensions are interdependent so that a modification in one leads to corresponding modifications in the others and, therefore, in the whole system. One can visualize such dimensions as circles or as a 'ring structure' (38). Usually the epistemological dimension of social systems does not receive formal attention. However, 'when a person enters a social system, he encounters a body of theory which more or less explicitly sets out not only the "way the world is," but "who we are," "what we are doing", and "what we should be doing"' (34). Consequently the 'theory' is a core dimension, and when it is essentially changed the organization is exposed to a critical disruption, to the extent that the change may affect: (1) its self-interpretation, (2) the definition of its goals, (3) the nature and scope of its operation, and (4) its transactions with the external environment.

TASK 2:
BLINDSPOTS OF CURRENT ORGANIZATION THEORY

It is a basic contention of this book that social systems whose design precludes substantive considerations, as a matter of course distort the language and concepts by which reality is apprehended. Accordingly it is to a substantive approach to organizations that our attention must now turn.

No significant change has occurred in the epistemological tenets of organizational analysis since Taylor. That is, organization theory has never critically confronted the epistemology inherent in the market system. The blind spots of current organizational theory can be characterized as follows:

1 / The concept of rationality prevailing in current organization theory is affected by strong ideological overtones. It leads to the identification of economizing behavior with human nature at large. Although the notion of economizing behavior seems self-evident, here it refers to any type of action undertaken by the individual when he is solely concerned with maximizing his economic gains.

2 / Current organization theory does not systematically distinguish between the substantive and the formal meaning of organization. This confusion obscures the fact that the formal economizing organization is a recent institutional innovation required by the imperative of capital accumulation and the enhancement of processing capabilities characteristic of the market system. The formal economizing organization cannot be considered a paradigm for all forms of organizations – the past, existing, and emerging ones.

3 / Current organization theory lacks a clear understanding of the role of symbolic interaction in interpersonal relationships at large.

4 / Current organization theory relies on a mechanomorphic view of man's productive activity. This is manifested by its failure to distinguish between labor and work.

To the extent that organization theorists continue to be unaware of these issues they succumb to a reductionist approach to social systems design. This reductionism requires that they see different types of social systems through a set of assumptions pertinent to only one of those types.

Each of these issues shall now be addressed in some detail.

The Notion of Rationality Re-examined
The state in which the notion of rationality finds itself in the field of organization theory illustrates its insufficient theoretical qualification. Herbert Simon's views of rationality presented in *Administrative Behavior* and other writings are still part of the conventional wisdom of the field. Rationality – as Simon's (1965) version goes – is absolute knowledge of consequences. Thus man hardly can be considered a rational being because comprehensive knowledge is beyond his capacity.[1] Corporate entities, however, such as the conventional organization, especially when they are computerized, deserve the attribute of rational to the extent that

they are less bounded than man in their calculative capability. Moreover, for Simon rationality in human conduct or decisions does not have an intrinsic qualitative content. Rather it denotes the instrumental character of such decisions in accomplishing objectives or ends. Therefore questions like the 'good' of man or society have no place in the area of rational debate. Rational man is unconcerned with the ethical nature of ends per se. He is a calculative being intent only on accurately finding adequate means to accomplish goals. The goals Simon considers are predominantly economizing in character.

Simon's contention has not been criticized on the intrinsic merits of his conception of rationality. Rather the prevailing criticism focuses on his argument that the organization cannot tolerate any kind of activity which is not attuned to the requirements of 'rationality.' Thus some organizational theorists argue for the individual's self-actualization within the frame of the organization and postulate a polarity between the rational man and the self-actualizing individual. The identification of rationality as calculability is taken for granted by both Simonists and anti-Simonists, as Chris Argyris's work illustrates. The assumption that self-actualization is not consistent with rational conduct is therefore widespread.[2]

It is my contention that both Simonists and their humanist critics beg the question of rationality. Until the rise of the market society the purposive type of reasoning which is concerned only with the means of achieving given goals had been a limited aspect of a broader concept of rationality. As explained elsewhere in this book, classically, the concept of rationality had always had ethical overtones, and to call a man (or a society) rational was to recognize his allegiance to an objective standard of values above any economizing imperatives. But Simon writes as though economizing were the sole criterion of rationality. There is not a single instance in his book in which he explicitly indicates the boundaries within which his concept is valid. Had Simon specified that his view was only valid in the world of pure economic pursuits his position would be more accurate. Unfortunately he fails to make such a qualification. In fact Simon tries to persuade the reader to believe that he is dealing with all that rationality could be. For example, he quarrels with Aristotle's concept of rationality which involves the inquiry into the 'good' of man and society and considers it 'limited' (47), as if he and the Greek philosopher were dealing with the same dimension of rationality.

The term 'rationality' as used by Simon has nothing whatsoever to do with the Aristotelian concept of rationality. Aristotle never supposed the market as the primal system of society and never thought that its psychological requirements should become the norms of social existence at large. He had a clear notion of the rationality of economizing behavior, but in his normative view of a good society such rationality should only incidentally impinge upon human existence. One could argue that Aristotle's concept of prudence contains a calculative ingredient. However, prudence in Aristotle's view is an ethical category, not purely a behavior of expedience. Thus the philosopher says: 'We cannot be prudent without being good' (*Nicomachean Ethics*, VI, XIII: 10). Aristotelian rationality and instrumental rationality belong to two different qualitative spheres of human existence. Aristotle's rationality cannot be criticized from Simon's perspective unless the author of *Administrative Behavior* indeed means that instrumental rationality is the only rationality conceivable, a position which is clearly truncated.

Historical Peculiarity of Economizing Organizations
The field of organization theory fails to understand the historical peculiarity of economizing organizations and their functions. The organization which constitutes the focus of attention of organization theory in *strictu sensu* is intrinsically tied to a society of an unprecedented type – the market society. As Marcel Mauss has pointed out, 'it is only our Western societies that quite recently turned man into an economic animal,'[3] i.e., into a creature who normally acts according to the utilitarian ethos inherent in today's formal organizations.

A distinction should be made between the substantive and the formal meaning of organization. This distinction is important for the same reasons which led Karl Polanyi to differentiate between the formal and the substantive meanings of the term economic. He says:

No society can exist without a system of some kind that ensures order in the production and distribution of goods. But that does not imply the existence of separate economic institutions; normally the economic order is merely a function of the social, in which it is contained. Neither under tribal, nor feudal, nor mercantile conditions was there ... a separate economic system in society. Nineteenth century society, in which economic activity was isolated and imputed to a distinctive economic motive, was, indeed, a singular departure ... Such an institutional pattern could not function unless society was somewhat subordinated to its requirement. A market economy can exist only in a market society. (Polanyi 1971a: 71)

Polanyi suggests that in non-market societies, economies existed in the substantive sense. In the market society, however, the term economic formally derives its sense from the assumption that means and resources are scarce and that they are to be optimized through choices which accurately meet economizing requirements. In non-market societies, scarcity of means does not constitute a formal principle for organizing production and for human choice at large, since the individual's livelihood is normally guaranteed by the effectiveness of overall cultural and social (not formal organizational) criteria of reciprocity, redistribution, and exchange. The economy, here, is embedded in the societal fabric and does not constitute a self-regulated system. In other words, in a non-market society human beings do not live under the threat of the 'economic whip.'[4]

For circumstances identical with those mentioned above, in non-market societies organizations generally are fields of experience of which no one is formally conscious. In these societies individuals have a compact, not a differentiated existence. They exist on substantive, not on formal, legal, or contractual grounds. For instance, a family in a primitive society is a substantive organization in that it would not function as a system unless some pattern existed in the relationships among those who constitute it, and between them and the external environment. The family in our society, to the extent that it still preserves some functions of the archaic family, shares its substantive organizational character. Yet thanks to the nature of contemporary global society, the family is becoming instead a formal organizational phenomenon. In another example demonstrated by Moreno (1934), groups on a playground constitute substantive organizations.

Unlike substantive organizations, formal organizations are grounded on calculation and as such they are contrived systems, deliberately created to maximize resources. As a concern of standard organization theory, they are social artifacts. In this sense, formal organizations of varying scope have existed in all societies although they have become an object of systematic study only in a late stage of history.

Indeed, in the most rudimentary societies people have been able to address themselves to problems of utilization of resources from the standpoint of calculated comparative advantages. For example, certain procedures for collecting fruits, hunting, fishing, building homes, and doing things of all sorts were recognized as being better than others from the standpoint of comparative results. Once an individual chooses one procedure rather than another he has indulged in a calculative sort of action. The desire for power inspired deliberate formal organizational frameworks in tribal societies, in ancient Europe, Greece, and Rome, and in specific institutions like the army and the church. Max Weber saw in these frameworks traits of what he called bureaucracy, or organization in the formal sense. But he was also aware that in such societies, these frameworks were delimited enclaves within the human life space. In these societies most of the human life space was kept available for social interaction free of formal organizational constraints. In other words, calculative types of action were incidental and often subsumed under rules of primary social interaction.

Weber understood that modern society is unprecedented to the extent that in it the formal organization (bureaucracy) became a primal social form, and its intrinsic calculative rationality became the dominant standard of rationality for human existence. Thanks to this circumstance modern society deserves to be labelled an organizational society, as it has been appropriately called.

The issues of man's life are diverse, and only a few of them belong essentially to the sphere of formal economizing organizations. In attempting to create and maximize resources necessary to his material well-being, the individual may indulge in mechanomorphic activities, which are those specific to the formal economizing organization. But operational, mechanical rules do not apply to the entire spectrum of human conduct.

Symbolic Interaction and Humanity
In every society man faces two problems: the problem of the meaning of his existence and the problem of his biological survival. A society is formed when it represents itself to its members as an expression of the order of the universe. Every society seems natural to its members to the extent that by adhering to its symbols and relying on its patterns they experience their existence as being in tune with such an order. In the words of Voegelin, 'every society has to cope with the problems of its pragmatic existence, and at the same time, it is concerned with the truth of its order' (Voegelin 1964: 2). In other words, in every society there exists on the one hand a range of actions symbolic in nature, actions mainly conditioned by the experience of meaning, and on the other, economizing activities which are mainly conditioned by the imperative of survival, of calculated maximization of resources. The criteria of each type of conduct are distinct and must not be confused. An economizing activity or an economizing social arrangement is valued for the empirical advan-

tages to which it leads. It is geared towards such advantages and not the experience of truth. Economizing activities are rewarding because of their extrinsic results, whereas symbolic interaction is intrinsically rewarding. The first type of activity is a means to an end, the second is an end in itself.[5]

In all primitive and archaic societies symbolic life has been predominant and has kept the patterns of economizing in a peripheral and subordinated condition. In primitive societies economizing activities are occasional, almost always restricted to instances when men face the problem of utilizing natural resources required by their pragmatic existence; and thus relations between men are never ordered solely by economizing criteria. In fact, anthropologists of a variety of theoretical persuasions have offered evidence that in pre-capitalist societies it is hard to discern exchange between individuals caused by a purely economic motivation.[6]

Before the market society there never existed a society in which the criterion of economizing became the standard of human existence. Current organization theory is essentially an expression of the market ideology. It is in the nature of this ideology to disregard the issues implied by symbolic interaction. For this reason conventional organizational theorists feel at ease in dealing with issues like trust, virtue, worth, love, self-actualization, and authenticity, in the realm of the economizing organization where, by their nature, they hardly belong.

Efforts to explain the nature of symbolic interaction are numerous. In this country one usually associates the issue with the works of the so-called Chicago school founded by George Herbert Mead. The topic, however, has also been a primary concern of authors whose theoretical orientations do not always coincide. These include Carl Jung, Ernst Cassirer, Georges Gurvitch, Eric Voegelin, Jürgen Habermas, Kenneth Burke, H.D. Duncan, Herbert Blumer, and many others. From their works, it seems possible to state some propositions which characterize the subject area of symbolic interaction theories.

1 / The symbolic interaction approach relies on the principle that there are multiple ways of knowing. Among other things it fundamentally questions the assumption that science in the scientistic sense is the only correct mode of knowledge. Cassirer explicitly affirms that science itself is one of several symbolic forms and that there is no reason to give it privileged status in relation to the others. Art, myth, religion, and history are forms of knowledge, entailing different kinds of experience, each of which is valid within the confines of its corresponding reality.[7]

2 / Symbolic interactionists assume that society is essentially social existence. The accent here is on *existence*, which cannot be explained by reifying categories such as 'forces,' 'structures,' or 'classes.' True existence, individual as well as social, is never a fact – a mere self-evident externality. It is an in-between,[8] a tension between the potential and the actual. Thus, social and individual existence cannot be explained by mechanomorphic categories such as those which plague the prevailing model of social science. Herbert Blumer observes that 'by and large sociologists do not study human society in terms of its acting units' but instead view it 'in terms of structure of organization' and 'treat social action as an expression of such structure of organization' with emphasis placed on 'structural categories, as social system, culture norms, values, social stratification, status positions, social roles, and institutional organizations' (Blumer 1962: 188–9). Characterizing the symbolic interaction approach, he says: '[It recognizes] the presence of organizations in human society and respects its importance. However, it sees and treats organizations differently. The difference is along two major lines. First, from the standpoint of symbolic interaction, the organization of human society is the framework inside of which social action takes place and is not the determinant of that action. Second, such organization and changes in it are the product of the activity of acting units and not of forces which leave such units out of account' (189).

In other words, the individual participates in the making of social reality and the character of this participation may differ from one individual to another. It can be an active one where the individual is a true existent (i.e., a subject) or it can be a purely reactive one. In the latter case, the individual loses the character of true existent and becomes a mere information processing system, as some computer scholars have it. It may happen that in certain circumstances social frameworks so heavily impinge upon individuals that they act as if they were completely molded by the social process. Conventional social science propositions would be correct is such a passive sort of reacting were to be equated with human nature itself. Its assumption is that the individual is a completely socialized being.[9]

3 / Symbolic interaction assumes that social reality makes itself intelligible to the individual through experiences free of formal operational constraints. Symbols are vehicles for the exchange of these experiences; that is, for the reciprocity of perspectives. In other words, such experiences of reality are socially exchanged or communicated through symbolic interaction, which necessarily entails intimate relations between indi-

viduals not mediated by superimposed patterns or rules of economizing. Symbolic interaction is a type of non-contrived, as opposed to contrived, communication. In purposive-rational systems like the conventional organization communications among individuals are not grounded on the free flow of the direct experience of reality, but are subsumed under a set of procedural, technical rules. The conventional organization would lose its *raison d'être* if it were to permit unobstructed symbolic interaction. Communications within conventional organizations are operational not expressive. In the realm of symbolic interaction there are no functional behaviors which are to be judged from the standpoint of instrumental strategies or technical rules, but intelligible and unintelligible actions or gestures defined from the vantage point of the reciprocity of perspectives. There is low tolerance for ambiguity in instrumental social interaction whereas there is high tolerance for ambiguity in symbolic interaction. 'One of the characteristics of symbols,' says Gurvitch (1971: 40), 'is that they reveal while veiling and veil while revealing and promote participation while preventing or restricting it, but nevertheless encouraging this participation.' Economizing activities are essentially bound to formal operational rules and, therefore, limit the scope of this type of intimacy in human transactions.

It is therefore evident that issues like love, trust, honesty, truth, and self-actualization should not be within the purview of the economizing organization, and such organizations must be distinguished from other types of social systems where those issues really belong. Economizing organizations make themselves intelligible rather through purposive-rational norms of conduct and communication. There are, however, other social systems where symbolic interaction is supposed to be the main grounds for intelligible interpersonal relationships.

Labor and Work

In all pre-market societies which display some degree of social differentiation there has always existed a clear distinction between high and low existentially ranked activities or occupations. Although the activities classified in one or the other category vary from one society to another, two assumptions seem to pervade such a distinction. First, activities of high existential rank are rather autonomously performed by the individual according to his desire for personal actualization. In exerting such activities the individual accomplishes something that in the eyes of other individuals is desirable as an end in itself. Second, activities which are of less than high rank are instead externally commanded by objective necessities and not by free personal deliberation. It is this second type of activity which constrains the individual to engage in toilsome efforts. High rank activities are not effortless, yet they are intrinsically rewarding.

It seems evident that a systematic distinction between labor and work can be conceptualized consistent with these assumptions. Labor is an exertion of effort subordinated to the objective necessities inherent in the process of production itself. Work is the exertion of efforts freely elicited by the individual in the pursuit of his personal actualization.

Such a distinction constitutes the background of Veblen's theory of leisure class. Labor, as defined above, has been so universally despised that those who do not need to labor in order to live make a point of stressing their condition through the practice of conspicuous consumption. Veblen's notion of conspicuous consumption may, however, preclude the full understanding of leisure.

In the market society the notion of leisure has been degraded by being made synonymous with idleness, hobby, and relaxation – connotations that leisure never had before. This is symptomatic of the value premises of the market pricing system in which labor has been transformed into the criterion par excellence of worth and merit. In a world of 'total labor' (Pieper 1963: 20), as implied by the market system, leisure naturally loses its previous character of being tantamount to a condition suited to the most serious endeavors in which a man can engage. Attempting to reconstitute the orginal meaning of leisure, Josef Pieper writes: 'Idleness, in the old sense of the word, so far from being synonymous with leisure, is more nearly the inner prerequisite which renders leisure impossible: it might be described as the utter absence of leisure, or the very opposite of leisure. Leisure is only possible when a man is at one with himself. Idleness and the incapacity for leisure correspond with one another. Leisure is the contrary of both' (40).

The inversion of the original meaning of leisure gradually accomplished through the process of ethical self-justification of the market system is an illustration of the bewilderment of Western civilization in its modern stage. 'Great subterranean changes in our scale of values' (23) took place in the last three centuries, whereby leisure lost its character as 'one of the foundations of Western culture' (20). This distortion was dictated by the value premises of the market system, where man feels that he is socially and even religiously justified 'to enjoy with a good conscience' only 'what he has acquired with toil and trouble' (33).

Veblen correctly points out that the existence of a leisure class is impossible without private property. This fact was well understood by

Aristotle who also specified that only those who had property of their own could be free. For him, to have property is a condition for a full-fledged, rational, leisurely life. He thus considers the slave as not a full rational being. Although this view is repugnant to our contemporary feelings, Aristotle is here only guilty of considering a circumstantial imperative as the indication of a dichotomy of essence between two sorts of human beings. As Leo Strauss points out: 'Aristotle took something for granted which we can no longer take for granted. He took for granted that every economy would be an economy of scarcity where the majority of men do not have leisure. We have discovered an economy of plenty. And, in an economy of plenty, it is no longer true that the majority of people have to be uneducated. This is a perfect reply to Aristotle as far as it goes. But we must see what precisely has changed. Not the principles of justice, they are the same. What has changed are the circumstances' (Strauss 1972: 231).

In fact, as long as the feasibility of an economy of plenty is inconceivable, it is correct to assume that in every differentiated political system only a minority could be free from the laboring condition, which is Aristotle's prerequisite for a rational and leisurely kind of life. Thus if we place our moral indignation against the justification of slavery in proper context, there is no argument against Aristotle.

As Arendt points out, 'the institution of slavery in antiquity was a device to exclude labor from the condition of man's life' (Arendt 1958: 74). This exclusion could only be feasible through the institutionalization of slavery, given the productive capabilities of that period of history. Reflecting on Aristotle's theory in its proper historical perspective, Arendt writes:

Aristotle, who argued this theory so explicitly, and then, on his deathbed, freed his slaves, may not have been so inconsistent as moderns are inclined to think. He denied not the slave's capacity to be human, but only the use of the word 'men' for members of the species mankind as long as they are totally subject to necessity. And it is true that the use of the word 'animal' in the concept *animal laborans*, as distinguished from the very questionable use of the same word in the term *animal rationale*, is fully justified. The *animal laborans* is indeed only one, at best the highest, of the animal species which populate the earth. (74–5)

The fact that words like reason, rationality, and leisure assume meanings in the market system that they originally did not convey is not accidental. The process of the institutional consolidation of the market system is inseparable from a process of deculturation of the Western mind whereby the original meanings of those words is obliterated. In particular, leisure and its implied qualitative distinction between labor and work were transformed in order to fit the term into the epistemological framework of the market system. In this market system labor became the source of all values and the *animal laborans* was elevated 'to the position traditionally held by the *animal rationale*' (Arendt 1958: 75).

How this transformation occurred is a very complex question which, incidentally, has been thoroughly discussed by W.A. Weisskopf (1957, 1971). This discussion focuses only on the psychocultural reasons for the 'sudden, spectacular rise of labor from the lowest, most despised position to the highest rank, as the most esteemed of all human activities' (Arendt 1958: 88). A summary of reasons follows.

First, the market system found optimal conditions to establish its command over social life during the so-called industrial revolution. Industry has now become a fundamental, constitutive piece of the market system. Industrial production relies on the laws of mechanics rather than on any singular personal dexterity. Industrial production effectively conditions man to comply with its operational requirements. In the process of manufacturing, labor is divided and thus the more the individual conforms to mechanical prescriptions in doing things, the better the overall results expected. Within these circumstances, and in order to accomplish its final outputs, personal skills are made subsidiary to mechanical designs. In other words, in such circumstances man is expected not to work and not to express himself freely concerning his assigned task, but he is expected to labor. Man is therefore essentially considered only as constitutive of a labor force.[10] The transformation of man into a laborer is a requirement of the mechanical design of production.

Second, the market system is a pricing system, and it needs objective standards to determine the equivalence of goods and services. Furthermore, insofar as the relationships between producers and consumers in the market are abstracted and simultaneously subsumed under a competitive process, profits and costs must be accurately calculated. Thus the individual participates in the production process, but only as an item of cost. Factors of production are priced, and thus the individual becomes merely a wage earner. In the market, as Blake remarked, 'souls of men are bought and sold.'[11] The transformation of man into a laborer is a requirement of production bookkeeping.

Third, the market system cannot function on purely technical and economic grounds. It could only become the leading social sector to the extent that the socialization process at large induced individuals to conform to its psychological requirements. Several scholars have studied the religious overtones of the ideology inherent in the market system and have indicated that such an ideology is not the contribution of a single person, but has resulted from concurring efforts of philosophers like Hobbes and Locke, religious reformers like Luther and Calvin, moralists like Bentham and others who have elaborated the theoretical background of the utilitarian ethos. The ultimate outcome of the endeavors of these men is the labor ethic, based on the postulate that labor is the cardinal criterion of value in the realms of individual and social existence. What in economics is known as the labor theory of value is only a particular aspect of the ideology which legitimizes the market-centered society.[12]

The choice of labor as a yardstick of value and human dignity at large was conditioned by the need to mitigate the cognitive dissonance generated by the rise of the market system. The old distinction between work and labor had to be undermined, otherwise inner conflicts in the human psyche would make the market system unworkable. Labor as a yardstick of value and human dignity is a psychocultural expedient used for minimizing cognitive dissonance and inner conflict.

The rudiment of a distinction between the two words is found in Alfred Marshall's *Principles of Economics*. Labor is here defined as 'any exertion of mind or body undergone partly or wholly with a view to some good other than the pleasure derived directly from the work.'[13] Although this statement satisfactorily clarifies the nature of labor, it is misleading at the end when the word 'work' is used. After citing Marshall's definition in *The Affluent Society*, Galbraith (1958: 264) accurately suggests that the formal distinction between labor and work has not played a role in economic theory. Galbraith seems to believe that the peculiar conditions of the 'affluent society' would require the distinction for the clarification of its problems. There is no doubt, however, that this distinction is theoretically relevant to a substantive approach to organizations.

CONCEPTUALIZATION OF A SUBSTANTIVE APPROACH TO ORGANIZATIONS

At this point, we must begin to grapple with the notion of organizational delimitation. The expression implies not simply that there are multiple types of organizations, but most importantly, that each of them belongs to distinct enclaves within the overall social fabric. Conventional formal organizations have so far constituted the main concern of contemporary organization theory. This focus has inhibited organizational theorists from systematically and accurately addressing themselves to the variety of social systems which constitute the macrosocial space. Overcoming this theoretical parochialism calls for a substantive approach to organizations. A substantive approach is characterized by the following considerations:

1 / The organization's boundaries should match its objectives. Accordingly, organizational delimitation is primarily concerned with the delimitation of the specific boundaries of the economizing organization. Tentatively, the economizing organization can be defined as a microsocial system which produces commodities under objective contractual norms, possesses operational devices for the maximization of limited resources, and uses quantitative criteria to evaluate the equivalence of goods and services. This definition means that economizing organizations, having requirements of their own which do not necessarily coincide with the requirements of the goodness of human existence at large, must be considered as belonging to a conceptually and pragmatically bounded enclave within the individual's life space.

2 / Human conduct within economizing organizations is necessarily subordinated to formal, superimposed operational constraints. Accordingly, 'administrative behavior' is inherently vexatious and incompatible with the full deployment of human potentialities.

3 / The economizing organization is only a particular case of several types of microsocial systems in which economizing functions are performed according to different scales of priorities. The relevance of 'administrative behavior' decreases when one moves from social systems designed for profit making toward social systems more suited for human actualization.

4 / A substantive approach to organizations is systematically concerned with means of eliminating unnecessary constraints on human activities in economizing organizations and social systems in general. In other words, it recognizes that by its very nature 'administrative behavior' is human activity under superimposed operational constraints. Nevertheless such an approach is concerned with finding viable means of minimizing and even eliminating dissatisfaction and with increasing the satisfaction of the members of economizing organizations.

5 / The sites where human beings are to come to grips with issues of their actualization, properly understood, have system requirements other than those of economizing settings. This systemic social differentia-

tion has been correctly envisioned by Hannah Arendt as a condition which enables individuals to excel in accomplishing the different deeds of their lifetime. She says: 'No activity can become excellent if the world does not provide a proper place for its exercise' (Arendt 1958: 49). To provide these proper places we must begin to formulate a typology of human concerns and of corresponding social systems where those concerns can properly be considered as issues of organization design.

As these five considerations indicate, a substantive approach to organizations resists being a tool of cognitive politics under any disguise.

CONCLUSION

From the analysis presented so far it is clear that organization theory must be reformulated on new epistemological grounds. From Taylor's time to the present organization theory – thanks to its persistent unexamined epistomological dimensions – has largely been an ideology of the market price system. It will survive only if it is transformed into a truly viable theory by becoming aware of its conceptual blind spots and by redefining itself on substantive grounds.

The following statements are offered as a set of tentative guidelines required for the reformulation of organization theory.

1 / Man has different kinds of needs, the satisfaction of which demands multiple types of social settings. It is possible not only to categorize such types of social systems but also to formulate the operational conditions peculiar to each of them.

2 / The market system meets only limited human needs and prescribes a particular type of social setting in which the individual is expected to perform according to rules of operational communication or purposive-instrumental criteria and as a laboring being. 'Administrative behavior' is, therefore, human conduct conditioned by economizing imperatives.

3 / Different categories of time and life-space correspond to different types of organizational settings. The category of time and life-space required by an economizing social setting is only a particular case among others to be discerned in the global ecology of human existence.

4 / Different cognitive systems belong to different organizational settings. The rules of cognition inherent in 'administrative behavior' constitute a particular case of a multidimensional epistemology of organizational settings design.

5 / Different social settings require distinct enclaves within the overall social fabric. There are, however, linkages which interrelate them. Such

linkages constitute a central concern of a substantive approach to social systems design.[14]

The scientific study of economizing organizations deals with structures leading to the effective utilization of physical resources and manpower. This scientific study of production, it is true, is concerned with personalities, but only to the extent that individual abilities and skills can be improved through training and effectively combined from the standpoint of desired outputs. Today's merging of the theory of economizing organizations and personality theory is a spurious union which veils a sinister intent. The only excuse for its advocates is, at best, their mistaken good faith.

In fact the proponents of 'scientific management,' like Taylor, and the positive-operationalists, like Herbert Simon, are closer to a valid organization theory than the 'humanists' theorists who misplace the notion of self-actualization.

In order to improve organization theory one should reformulate Taylor and Simon. The effort of these men in trying to discover effective structures that economizing organizations should have in order to reach their goals is meaningful. They were concerned essentially with technical questions and most of what they have said still constitutes at least a partial foundation on which one can continue further theory building. There is much to be considered valid in Taylor's effort to formulate the foundations of a science of production. Simon was also basically correct in his attempt at a clarification of the decision-making process, as it characteristically takes place within the confines of the economizing organization. However, when they overlooked the boundaries of the rules of cognition inherent in economizing organizations, their psychology objectively became a bad faith psychology because they unconsciously transformed them into rules of cognition supposedly valid for human nature in general.

Taylor and Simon still deserve to be re-examined from the perspective of a theory of economizing organizations, but a theory contained within its proper bounds. The organization theory they envisioned deals with human activities operationally useful to economize resources. The error of Taylor consisted in over-expanding the logic of such specific activities. For Taylor every act of human life should be focused from the standpoint of scientific management. He seems to be unconcerned with the role of primary (symbolic) social interaction, a realm of association which has nothing to do with calculability and maximization. In the last analysis Simon equates rationality with the 'reckoning of consequences' and therefore identifies the psychological requirements of the market system with human nature at large.

The problem of doctrinaire viewpoints of these sorts is not only that they are theoretically unsound, but also that through the practice of the politics of cognition they are utilized to construct the social reality of the common citizen.

There is no point in dismissing the scientific study of economizing social settings. Society at large cannot subsist without them. Their design and operation constitute a technical problem of a peculiar kind. However, such a domain is only a part of what in the next chapter will be conceptualized as a theory of social systems delimitation.

7

Theory of Social Systems Delimitation: A Paradigmatic Statement

The model of social systems analysis and design which now predominates in the fields of administration, political science, economics, and social science at large, is unidimensional. It reflects the modern paradigm which largely assumes the market as the cardinal category for ordering personal and social affairs. In this chapter I will begin to delineate a multidimensional model for social systems analysis and design in which the market is considered a legitimate and necessary, but limited and regulated social enclave. Such a model is reflective of what I call the para-economic paradigm.

Central to this multidimensional model is the notion of organizational delimitation. This notion implies (1) a view of society as constituted by a variety of enclaves (of which the market is but one) in which man engages in distinctly differing, yet truly integrative, types of substantive pursuits, and (2) a societal governance system able to formulate and implement allocative policies and decisions required for optimal transactions between those social enclaves. This chapter focuses primarily on the first implication. The second implication is examined in chapters 8 and 9.

The accompanying diagram sets forth the major dimensions of the para-economic paradigm. The categories of the paradigm (in italics) must be considered as heuristic constructs in the Weberian sense. No situation in social life is expected to coincide with these ideal types. In the concrete world only mixed social systems exist.

An explanation of some specific details of the paradigm is now in order.

PERSONAL AND COMMUNITY ORIENTATION

In the social world envisioned by the paradigm there are places for individual production and actualization free from superimposed pre-

THE PARA-ECONOMIC PARADIGM

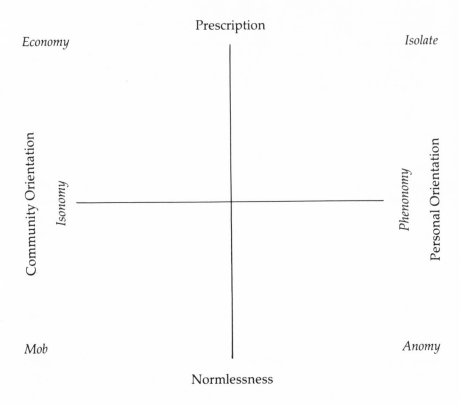

scriptions. This kind of personal action can occur either in small exclusive settings or in communities of moderate size. In these alternative settings, true personal choice is possible. One must bear in mind that in the epistemological framework of the para-economic paradigm, personal choice does not have the same connotations of the word 'choice' in the field of current policy sciences and especially as employed by 'public choice' theorists.[1] These theorists would see personal choice where, from the viewpoint of the paradigm, there is none. They reduce the individual or the citizen to a utility-maximizing actor, permanently engaged in trading activities. The choice exercised by this actor does not confront the market, but assumes that the individual is completely included in it. His nature is defined by the requirements of the market. 'Public choice theory' as well as 'administrative theory' is predicated on a unidimensional model of man which envisions social space as horizontal and flat: wherever the individual goes, he never leaves the market.

In contrast, first and foremost, the para-economic paradigm assumes that the market is an enclave within a multicentric social reality where there are discontinuities of several sorts, multiple substantive criteria of personal life, and a variety of designs of interpersonal relations. Second, in this social reality the individual is only incidentally a utility maximizer; basically he strives for ordering his existence according to his needs for personal actualization. Third, the individual with access to alternative social spaces is not compelled to total conformity to the market price system. He is granted opportunities to 'work' or even 'beat' the market system, creating and participating in a variety of social settings which differ one from the other in nature. In summary, the space depicted by the paradigm is one in which the individual can properly act, rather than merely behave so as to meet the expectations of a market-dominated social reality.

One can seldom 'integrate' personal actualization with utility maximization in the strict economic sense. Whenever the two are seriously considered as fundamental imperatives of individual and social life, one must delimit enclaves where each can be cogently accomplished. Utility maximization is incidental in systems designed for personal actualization, and, conversely, personal actualization is incidental in systems designed for utility maximization. Social systems design is a multidimensional science. Living in accord with the para-economic paradigm is a multidimensional art.

Organizational delimitation is therefore a systematic attempt to overcome the continuous process of unidimensionalization of individual and collective life. Unidimensionalization is a specific type of socialization by which the individual deeply internalizes the market ethos, and acts as if this ethos were the overarching normative standard of the entire spectrum of his interpersonal relations.[2] This process is characteristic of the market-centered society in the peculiar institutional form it has assumed in advanced industrial countries.[3]

The process of unidimensionalization has been treated by a number of writers of varying philosophical persuasions. I do not intend to belabor this point. A book by Philip Slater, however, should be noted: *The Pursuit of Loneliness: American Culture at the Breaking Point (1971)*. Slater investigates the psychological and social outcomes of unidimensionalization. In the social world he describes, 'public relations, television drama and life become indistinguishable' (19) and the individual is systematically taught to misexpress his emotions. Such a world engenders what Slater calls the 'warping of human emotionality' (p. 3). He sees the process of uni-dimensionalization of American society approaching a breaking point. In the past, Americans had room available to design existential arrange-ments more suited to their personal choices. Thus he states:

In the past, as so many have pointed out, there were in our society many cases in which one could take refuge from the frenzied invidiousness of our economic system – institutions such as the extended family and the stable neighborhood in which one could take pleasure from something other than winning a symbolic victory over one of his fellows. But those have disappeared one by one, leaving the individual more and more in a situation in which he must try to satisfy his affiliative and invidious needs in the same place. This made the balance a more brittle one – the appeal of cooperative living more seductive, and the need to suppress our longing for it more acute. p. 6.

An art of social systems design concerned with human actualization in its own right, as well as with effectiveness in production of good and delivery of services, has to postulate a variety of organizational settings where these different objectives are likely to be accomplished. The unqualified statement that the concern for people can be reconciled with the concern for production of commodities can only be justified on the basis of a unidimensional approach to organizations. This is precisely the fallacy characteristic of current trends in administrative thought and practice. Examples include designations such as 'Theory x versus Theory y,' 'managerial grid,' and 'organizational development.' Instead of claim-ing the possibility of a total integration of individual and organizational goals, the paradigm introduced here indicates that human actualization is a complex endeavor. It can never be undertaken in a single type of organization. As a jobholder the individual is usually compelled to act under superordinated prescriptions. In different degrees, however, the individual has a variety of needs. For instance, he needs to share in the community as well as to engage in ventures which express his unique-ness. The settings adequate for the satisfaction of these needs, while largely unstructured, are to some extent shaped by prescriptions either reached by consensus or freely self-imposed. The discussion of the inter-nal governance terms peculiar to differing social spaces will be undertaken in a later stage of this analysis. It is important, now, to further delineate the overall paradigm under examination.

PRESCRIPTION VERSUS NORMLESSNESS

In order to get any job done operational prescriptions must be met. The more the job is of an economizing nature, the less its operational prescriptions leave its holders room for personal actualization, because a minimal opportunity exists for personal choice in the sense in which it has been discussed above. This contradiction between the needs of the individual and the requirements of the economizing organization cannot be resolved by any behavioralistic or so-called humanistic procedure. Production of goods and the delivery of services under the imperative of maximization of the net balance of costs and benefits calls for types of organizations where there is, obviously, low tolerance for personal actualization. In fact, the word 'behavior' in this context conveys what people are expected to do as jobholders in economies. Thus, as it was said earlier, administrative behavior is human activity under superimposed and formal operational prescriptions. The unqualified use of the expression 'administrative behavior' is itself an indication of the unidimensional character of current organizational theory and practice. This theory systematically overlooks the fact that administrative behavior is a category of conformity to superimposed and formal prescriptions. The more human action is considered administrative, the less it is an expression of personal actualization.

As they characteristically function in the market-centered society, economies are to a certain extent threat systems, provided with means to compel their members to comply with their operational prescriptions. They say to the individual: conform to prescriptions of performance or leave. Administrative behavior is a psychological syndrome inherent in economies and threat systems in general. The problem, however, with the current type of unidimensional organizational theory and practice is that it assumes that administrative behavior is identical to human nature. This erroneous assumption is sometimes made in crude terms. For instance, in one typical behavioral textbook one reads that 'the organization is believed to have, on a large scale, all the qualities of the individual' (Rush 1969: 8). Under the pressures of the market system it is not surprising that the average individual is confused about both the nature of humanness and personal actualization. Current 'administrative theory' legitimizes the ongoing process of over-organization and depersonalization of the individual within the market system of an advanced industrial type.

This double process of over-organization and depersonalization might be characterized as follows:

The phenomenon of over-organization in American society has been investigated by a number of specialists. Over-organization occurs with the transformation of the entire society into an operationalized universe in which the individual is always expected to live as an actor with a prescribed role.

In an over-organized social system the individual is deprived of a truly private place as well as of time, two conditions for a creative personal life. 'Solitude,' says Marcuse, 'the very condition which sustained the individual against and beyond his society, has become technically impossible' (1966: 70). One can easily substantiate this statement. In one insightful essay Mordecai Roshwald provides a number of striking illustrations. He refers to the standardization of emotionality which results from the widespread use of cards available in shops for customers to express their feelings on occasions such as birthdays, anniversaries, weddings, death, illnesses, or for sympathy; the use of cheerleaders to arouse football spectators; and cues for laughing and applauding as signals to audiences of programs. Education also has not escaped the process of over-organization. In general its goal is largely to enable people to become jobholders in the market system. Students in high schools and under-graduate colleges are submitted to uniform procedures of teaching and evaluation which hardly encourage them to be creative and to develop their sensitivity to the complexities of issues on which they are directed to focus.[4] Continuously trapped within requirements of orderliness and organization, the individual ends up yielding to a prescribed view of reality.

Obviously over-organization increases the depersonalization of the individual. One finds in the works of Erving Goffman, for instance, abundant clinical material on the institutionalized pressures of depersonalization. One of Goffman's conclusions is that a predominant model of interpersonal transaction in this society is 'impression management,' or the practice of systematic deception among people. Goffman's findings give empirical support to the affirmation that in an over-organized society the individual loses personal identity as he is induced to internalize a prescribed identity required by the roles he is expected to perform.[5]

A multidimensional art of social systems design cannot overlook the psychological effects of operational prescriptions. It does not seek to eliminate those prescriptions from the social world, because they are indispensable to the maintenance and development of the supporting system of any collectivity. This art is, however, concerned with delimiting the enclaves where those prescriptions obtain and where they may even be legitimately superimposed upon the individual. In social systems intended to maximize personal actualization, prescriptions are not eliminated. They are, however, minimal and never established without the full consent of the individuals involved. These systems are flexible enough to enhance the individuals' sense of order and commitment to goals without transforming them into passive role incumbents. Complete elimination of prescriptions and norms is incompatible with meaningful human actualization within the social world. Accordingly, events classified under the categories of *mob* and *anomy* essentially jeopardize the viability of the overall social fabric.

In the context of this chapter, the designer of a social system is not considered as a sort of patron or Pygmalion who shapes a setting and tells its members how they should live in it. Rather, the designer is conceived as an agent able to facilitate the development of initiatives freely generated by individuals and to coalesce them in the form of effective configurations. In this capacity he may perform some of the roles which characterize Donald Schon's network manager,[6] such as facilitator, systems negotiator, 'underground' manager, maneuverer, or broker. One can add to these other roles such as team builder, group dynamics expert, group therapists as represented, for instance, by Ira Progoff, and the role of the space designer as described by Fred Steele.[7]

Finally, it must be pointed out that enclaves as conceptualized in the paradigm are not supposed to exist in segregated parts of the physical space. Economies, isonomies, phenonomies, and their mixed forms are characterized by their specific life styles and eventually may be found in physical vicinity.

CONCEPTUALIZATION OF DELIMITATIONAL CATEGORIES

A conceptualization of each category represented in the paradigm is now in order.

Anomy and Mob
The presence of the categories of *anomy* and *mob* in the paradigm is required by the logic of its dimensions. *Anomy* is conceptualized as a

limited situation in which personal and social life vanishes. The term *anomy* (in French, *anomie*), originally coined by the French sociologist Émile Durkheim, defines a condition in which individuals subsist at the fringe of the social system. They are normless and rootless, uncommitted to operational prescriptions, but unable to shape their life according to a personal project. They are the 'beats,' the marginals, the excluded (who occasionally assume the condition of the wanderer youth or unconventional adult in search of their identity or new experiences) some criminals, drug addicts, drunkards and beggars, the indigent, and the mentally crippled.

The anomic individual is unable to create a social setting for himself and simultaneously to comply with the operational prescriptions of organizations required for his livelihood. He has to be assisted, protected or controlled by institutions like the Salvation Army, asylums, reformatories, hospitals, and prisons. Yet the design of settings for anomic individuals has to meet specific requirements. Articulators and caretakers of these systems should realize that their task involves means and skills suited to the objectives at hand. In this sense *anomy* fits into the delimitational frame. One reason why the above-mentioned institutions usually aggravate the anomic condition of people under their care is that their design and their management are not systematically understood as pertinent to a specific social enclave. The clients of these institutions, which in fact constitute a living testimony to the prevailing social malaise, are defined according to the pervasive operational assumptions of the society at large. This, in itself, inhibits their agents' understanding the nature of their assignment and the qualifications they are supposed to have. A great number of settings assigned to deal with anomic individuals are being widely experimented with today. The design and implementation of such settings, as for instance reported by J.M.N. Query (1973), S.B. Sarason (1974), and others, involves a specific expertise which now is still in a very incipient stage. If a delimitation of the market is ever feasible, then the structure, functions, and assumptions of these institutions will be radically different from the ones now prevailing. Accordingly, *anomy* deserves to be considered a category of organizational delimitation.

In the paradigm *anomy* refers to normless individuals who lack a sense of relationship with others. *Mob* refers to normless collectivities of individuals who lack a sense of social order. A society may eventually become prone to be upset by *mobs* when it loses representativeness and meaning for its members.

Economy

In general terms an economy is a highly prescriptive organizational setting established to produce goods and / or deliver services, and possessing the following characteristics:

1 / It serves customers and/or clients who have at best an indirect influence on the planning and execution of its activities.

2 / Its survival is a function of its effectiveness in producing goods and delivering services to customers and clients. Thus the effectiveness of an economy can be objectively evaluated either in terms of profits and/or cost-benefit ratios involving more than mere consideration of direct profits.

3 / It can and usually must assume large dimensions in size (expressed by the amount of personnel, offices, physical facilities, and so forth) and complexity (expressed by the diversity of activities, duties, transactions with the environment, and so forth).

4 / Its members are jobholders and are primarily evaluated as such. Professional qualifications for performing jobs dictate their hiring, firing, retention, promotion, and decisions on advancement.

5 / Information circulates asymmetrically among its members as well as between the economy as a setting and the public. This means that people at various hierarchical levels of the system condition the delivery of information to personal and/or corporate interests. Such a pervasive condition of economies in general is the main factor of the 'iron law of oligarchy,' Parkinson's law, the Peter principle, misplacement of goals, and so forth.

The above five characteristics are assumed to be common to all economies: monopolies, competitive firms, non-profit organizations, and bureaus. Obviously each of these four types of economies can be examined in terms of their peculiarities as well as their commonalities. But a detailed analysis of these economies is not essential to the objectives of this study. Suffice it to say that although monopolies, competitive firms, and even non-profit pursuits draw their revenue from per unit production, bureaus operate on a budget basis deriving their income at least partially from grants, gifts, direct funding, and appropriations.

The market tends to become an all-inclusive category for ordering individual and social life. In the market-centered society economies are free to shape the minds of their members and lives of their citizens in general. Thus a market-centered political and administrative theory, as is characteristic of the one which now prevails and is widely taught, assumes that the criteria of effective performance in the exchange between individuals and economies epitomize the essence of human nature. The predominance of economies and their characteristic administrative behavior are considered by the market-centered political, economic, and administrative theorists as axiomatic. To illustrate this, a comment on a passage from Victor Thompson's *Modern Organization* (1966) is appropriate.

Thompson asserts that 'success within our society means, for the most part, progression up an organizational hierarchy' (21). He seems to focus upon this issue from a viewpoint similar to Emile Durkheim's oversocialized conceptualization of human normality. Ultimately for Durkheim, and apparently also for Thompson, the criteria of normality and morality in human life are inherent in the social system. This orientation explains the value-neutral overtone of the following statement by Thompson:

Since a monocratic institution cannot admit the legitimacy of conflicts, the legitimacy of divergent goals and interests, much effort is spent securing the appearance of consensus and agreement – securing a 'smooth-running organization.' The modern organization wants converts as much as it wants workers. It is concerned with the thoughts of its members as well as their actions and with the thoughts of its public about the thoughts and actions of its members. Consequently, it is concerned with its members' total lives, with what they think and do away from work as well as at work. (20–1)

One must say that Thompson's *Modern Organization* offers a very accurate description of economizing behavior. However, his complete lack of concern with organizational delimitation lends a unidimensional character to some of his views. For instance, in explaining his concept of bureausis, he construes the psychological patterns required by bureaus as standards of human health. Nonconformity with those patterns (*bureausis*) is, for Thompson, a reflection of an immature personality (p. 24). This view ultimately legitimizes the process of unidimensionalization as previously described.

Recently a number of writers and practitioners have developed views in apparently direct opposition to those of Thompson. They advocate, without qualification, non-hierarchical organization, participative management, and sometimes the total elimination of bureaucracy. One of them has predicted the disappearance of bureaucracy within the next twenty or fifty years. In general the views of these authors are expressed in sweeping terms. What they fundamentally lack is a coherent and systematic view of organizational delimitation.

To conclude this section, it should be pointed out that participative management, involving non-hierarchical interpersonal transactions, is an issue largely strange to market-centered economizing settings. Since at this point in history it is inconceivable that any society will ever be completely able to disregard economizing activities, some degree of hierarchy and coercion will always be required for ordering human affairs at large. Within their pertinent enclaves, bureaucratized economies can become more productive for their members and citizens in general.

Isonomy
In general isonomy can be defined as a setting in which its members are peers. The *polis*, as conceived by Aristotle, was an isonomy, an association of equals constituted 'for the sake of a good life' (*Politics*, I, ii, 125b, & 8). The use of this term, however, does not imply any nostalgic craving for a return to a past period, but serves only to call attention to possible forms of present-day egalitarian social settings. The following are the main characteristics of isonomy:

1 / Its essential goal is to allow the actualization of its members free from superimposed prescriptions. Thus prescriptions are minimal and, when unavoidable, are established by consensus. Individuals are expected to engage in interpersonal relationships as long as they enhance the good life of the whole.

2 / It is largely self-gratifying in the sense that in it freely associated individuals accomplish activities that are rewarding in themselves. People do not earn their living in an isonomy; rather, they engage in a give-and-take, generous kind of social transaction.

3 / Its activities are undertaken primarily as vocations, not as jobs. In isonomies people work; they do not labor. In other words, their basic reward is the accomplishment of their works' intrinsic objective, not the revenue eventually drawn from their doings. Thus utility maximization is incidental to the individual's basic concerns.

4 / Its decision- and policy-making system is all-inclusive. There is no differentiation between the leadership or the management and the subordinates. Thus an isonomy would lose its character if its members dichotomized themselves as 'we' and 'they,' the latter understood as those who make decisions or set policies. Isonomy is conceived as a true community in which authority is assigned by common deliberation. Authority continuously moves from person to person according to the

nature of the issues and problems at stake and the qualifications of people to deal with them. The suffix *nomos* is particularly indicative of the fact that in this type of association there is no fixed and exclusive ruling agency as the suffixes *archy* and *cracy* – in monarchy, oligarchy, and democracy – would suggest. An isonomy is not a democracy. This leads to its fifth characteristic.

5 / Its effectiveness requires that primary interpersonal relations prevail among its members. If its size increases beyond a certain optimal point so that secondary or categoric relationships among people emerge and develop, the isonomy will necessarily decay, and eventually will transform itself into a democracy, oligarchy, or bureaucracy.

Isonomy is increasingly becoming a part of the present social world. One may not find a full materialization of the concept, which after all serves only a heuristic purpose. But everyone can think of tentative isonomic settings already working in this country, as, for instance, PTAS (parent-teacher associations), student and minority associations, urban communities, worker-owned firms, some artistic and religious associations, local associations of consumers, groups and citizens concerned with community issues and problems, and many other recently constituted arrangements in which ultimately people are searching for styles of life which transcend the dominant normative standards of society at large. Types of neighborhood governments and community development corporations envisioned by practitioners like Saul Alinsky,[8] Milton Kotler (1969), Simon S. Gottschalk (1973), Will McWhinney (1973), and others, and designs implied in the approach to social reform and reconstruction suggested by Lewis Mumford, Paul and Percival Goodman, E.F. Schumacher, Ivan Illich, H.R. Shapiro, and others, have a large isonomic component. The tools necessary for such a reconstruction as Illich calls 'convivial' now constitute a systematic subject-area of increasing interest to social reformers and technologists. That a technology of 'convivial' tools already exists is demonstrated by publications like *Alternative Sources of Energy* (Sandy Eccli, et al. ed), Victor Papanek's *Design for the Real World*, and others.

Phenonomy
This is a breakthrough social setting, either sporadic or more or less stable, initiated and run by an individual or a small group. It entails for its members the maximum degree of personal choice and a minimum degree of subordination to formal operational presciption. A phenonomy has the following main characteristics:

1 / It is constituted as a setting necessary for people to release their creativity in autonomously selected forms and ways, and is part of the expressive endeavor (in Greek *phaineim* means 'to show') which mobilizes the creative efforts of a small group or a single individual.

2 / Its members are engaged only in self-selected works. This means that they usually are extremely busy and seriously committed to accomplishing what they personally deem relevant. It is important to emphasize that self-motivated tasks are more often than not the most costly in effort. In order to successfully carry them out individuals have to develop their own schedules and operational rules and never allow themselves to act capriciously.

3 / Although the output of activities undertaken in phenonomies can eventually be marketable, economizing criteria are incidental to the orientation of their members. Phenonomies are social settings guarded against the penetration of the market. This point must not be missed if one wants to understand the nature of a phenonomy. In fact phenonomies confront or 'beat' the market system. The office that the upwardly mobile employee or overly busy manager maintains in his home in order to accomplish the toilsome activities required by the job at night and on week-ends is not a phenonomy.

4 / Although concerned with his or her uniqueness, the member of a phenonomy is socially mindful. Actually his or her choice is not to withdraw from society at large, but to sensitize other individuals for possible experiences that they are able to share or appreciate.

A number of people are normally involved in pursuits which qualify as phenonomies. This is the case, for instance, with the gifted housewife and husband who systematically reserve a spot in their home to engage in creative pursuits like the design and production of rugs, pottery, or paintings, as well as with the workshops of free lance artists, writers, journalists, craftsmen, inventors, and so forth. An illustration of particularly successful phenonomies is the one in which Will and Ariel Durant have been accomplishing their lifelong projected series of historical and philosophical essays, and the artistic venture of Simon Rodia, the title setter and general repair man who built the justly famous towers in Watts (Los Angeles).

The Isolate
While the anomic individual and the members of a *mob* are normless, the isolated actor represented in the paradigm is overcommitted to a norm unique to himself. For a variety of reasons, the *isolate* deems the social

world at large rather unmanageable and hopeless. But in spite of his total inward confrontation with the social system at large, he finds a spot in which he consistently can live according to his peculiar and rigid belief system. This is not the case with the anomic individual who fails in developing a personal belief system as well as in conforming to the overall set of social standards. Isolates may be considered clinical cases of paranoia, but not necessarily. A number of them are non-involved jobholders and citizens who systematically hide from others their inner personal convictions.

Thus conceptualized, the para-economic paradigm constitutes the referent for a new approach to social systems design and the new science of organizations. These matters will be examined in the following two chapters.

8

The Law of Requisite Adequacy and Social Systems Design

Much of what constitutes the para-economic paradigm is a categorization of basic trends of the emerging post-industrial society. This is not to say that the para-economic paradigm implies an evolutionary conception of the historic and social process. Current scenarios and explanations of the post-industrial society are still largely captured by thinking patterns predicated upon nineteenth-century serialist theories. In contradistinction, the para-economic paradigm does not envision post-industrial society as the necessary unfolding of a market-centered society. There is, of course, no guarantee that the literal extrapolation of this society's intrinsic trends will lead to a multicentric society as categorized by the para-economic paradigm. Rather it is more likely that such an extrapolation will aggravate the malaise which afflicts contemporary men, some aspects of which have been pointed out in this book. Therefore, the post-industrial society envisioned by the para-economic paradigm can only come into being as a result of confrontive endeavors on the part of actors whose personal project is to resist the intrinsic trends of the market-centered society. Yet the objective of the para-economic paradigm is not to obliterate the market mechanism, but to save only the unprecedented capabilities it has created, albeit for wrong reasons. Properly delimited, the market does serve the goals of a multidimensional model of human existence in a multicentric society.

The multicentric society is a deliberate undertaking. It implies design and implementation of a new kind of state empowered to formulate and enforce allocative policies supportive not only of market-oriented pursuits, but of social settings suited for personal actualization, convivial relationships, and community activities of citizens as well. Such a society also requires the initiatives of citizens, who are stepping out of the market-centered society at their own responsibility and risk.

The para-economic paradigm assumes that post-industrial life designs are immediately possible, both in centric and peripheral nations. It is a 'do-it-yourself' approach to the post-industrial society. For many individuals the post-industrial society is not a future stage, but to a significant degree an objective possibility within their reach.[1] Today, the delimitative model captures in a conceptual fashion the way of life sought after by many people in a number of places. Unfortunately the inchoate social systems which these individuals are creating by trial and error have not yet become the thrust of a systematic and disciplined effort of theory building in academic milieux.

Expertise is already available to design and monitor economizing social systems. There is less than sufficient expertise to design and monitor social systems in which economizing activities are, at best, of incidental character. As a result dominant organizational scholarship is scarcely providing the knowledge needed to overcome the social plight of contemporary man. One of the objectives of the para-economic paradigm is to formulate guidelines of a new organizational scholarship attuned to the operative realities of a multicentric society.

One fundamental topic of the new science of organizations is what I call the law of requisite adequacy. As a matter of fact, I would prefer to call it the law of requisite variety. But this expression has been used by W. Ross Ashby (1968) to analyze rather physical and biological systems. The question of social systems delimitation is alien to Ashby's conception of the law of requisite variety. Social systems delimitation postulates a variety of differentiated settings as vital imperatives of sound human associated life, i.e., it implies that men's actualization is thwarted when they are coerced to conform to a society pre-emted by the market or any other social enclave. Specifically, the law of requisite adequacy states that a variety of social systems is an essential qualification of any society which is responsive to its members' basic needs of actualization, and that each of these social systems prescribes design requisites of its own.[2]

Mary Parker Follet showed a sensitivity to a partial feature of this topic when she directed managerial attention to the 'law of the situation.' Her concern was to free management from arbitrariness and 'bossism,' and to envision it as a depersonalized process of giving and receiving orders. In Follet's view of organization, authority and consent are 'parts of an inclusive situation' (Follet 1973: 33) in which managers yield to norms induced from concrete circumstances. Follet focuses mainly upon the process of 'giving orders.' The law of requisite adequacy has a wider scope. It also suggests that although systems requisites can be generally conceptualized, for the systems designer they are rather a practical issue, i.e., the outcomes of concrete participative observation involving oneself and one's clients. I shall illustrate the meaning of this law through a brief discussion of some salient dimensions of social systems, namely technology, size, space, cognition, and time. At the present juncture of my research I can only formulate conjectural and impressionistic statements on these matters.

TECHNOLOGY

Only a brief comment on this dimension of social systems seems in order here since this topic has been rather thoroughly studied by conventional organizational scholars. There is an accumulated literature on this topic from which designers of confrontive social systems can draw useful and relevant knowledge. It is generally acknowledged that technology is an essential part of the supportive structure of any social system and exists in the body of operational procedures and tools through which things get done. Thus there is no social system without a technology, whether it be, for instance, a church, a prison, a family, a neighborhood, a school, or a factory. When called upon, the designer ought to include as a central aspect of his analysis examination of whether the technology used by the social system enhances, or hinders, accomplishment of its goal. Meeting this imperative involves complex analytic work, which the designer should undertake in close collaboration with his clients. Much of the success of what in the domain of standard organizational scholarship is known as *socio-technical systems* results from the systematic attention its representatives have given to the fit between a social system's technology and the system's specific goals. This skill is well developed. Moreover it has a general scope and should be assimilated and expanded by designers of confrontive social systems.[3]

SIZE

By contrast, there is relatively little contemporary scholarship which systematically addresses itself to the matter of size. This is not to say that the issue has been ignored. On the contrary, there is a legacy of knowledge on the impingement of size (i.e., number of people) of social settings upon their effectiveness and the character of interpersonal relations of their members, which deserves re-elaboration and systematization. Yet, so far, the size of social systems has scarcely become one of the central concerns of institution builders and social systems designers. The endurance of organizations or social systems is more important than what is today currently named 'organizational development,' which, more often than not, considers the size of existing organizations either as a given or as a less than relevant issue.

In a cultural environment like ours in which the assumption 'the bigger the better' is pervasive there is a need to emphatically point out that the effectiveness of a social setting in accomplishing its goals and making optimum use of its resources does not necessarily imply an increase in size. The principle 'the bigger the better' very often leads to deceptive interpersonal relations, Parkinson's law syndrome, unnecessary redundancy, and ultimately to wasteful social systems with poor self-sustaining capability. We need to learn the art of designing social settings which are capable of enduring.

The size of social settings has been a subject investigated by reformers and political theorists. Plato in a meticulous fashion stated that the good polity should have 5,040 citizens (heads of family). Aristotle avoided the arithmetical precision of his master but was identically aware that limits should be imposed on the size of the citizenry as one of the conditions of a good polity. Reflections on the subject are also found in the works of Montesquieu and Rousseau. In the Federalist papers (no. 14) James Madison addresses the issue of size as a potentially prohibitory factor for the functioning of the principle of representation in the Union. And in a broad sociological framework the German sociologist Georg Simmel (1950) focuses upon the quantitative aspects of social relations. He is a pioneering figure of what is now known as 'group dynamics.'

A contemporary scholar, Robert Dahl, has underscored the relevance of the size dimension of political systems. In his book *After the Revolution?*, the theme is highlighted. It is more thoroughly investigated in *Size and Democracy*, which Dahl wrote in collaboration with Edward R. Tufte. Extremely relevant are also *The Breakdown of Nations* (1957) and *Overdeveloped Nations* (1977), two seminal volumes in which Leopold Kohr presents his size theory of social and economic development. Significant findings are presented in these books worth consideration by anyone involved in a research effort aimed at developing skills for dealing with questions of scale in social settings. Drawing from these and other sources I will venture to propose three tentative statements.

First, *the capability of a social setting to endure and to respond effectively to the needs of its members imposes minimum or maximal limits on its size.*[4] In other words, each social setting has a concrete size limit below or above which it loses capability to accomplish its goals effectively (e.g., to extract and process resources) and to elicit from its members the minimum of consensus it needs to preserve itself.

Second, *no general procedure can be formulated to determine precisely before the fact the size limit of a social setting; the question of size is always a concrete problem to be solved through ad hoc investigation in context.* In other words, it is possible to determine accurately the size limit of a social setting. But this task is a question which involves not only *technical competence* but a *trained sensitivity* to the mutual implications of context and design.

Third, *the intensity of face-to-face relations between the members of a social setting tends to decline in direct proportion to the increase of its size.* A corollary of this postulate is that when the intensity of face-to-face interpersonal relations is deemed fundamental for the accomplishment of a goal, small rather than large-scale settings are appropriate.[5]

There is no general rule for determining the size of economies. For instance, economies of isonomic character, i.e., certain types of co-operatives and enterprises in which management and ownership are collective prescribe rather moderate sizes. Yet when division of labor, impersonality, and specialization are indispensable for economies to compete successfully in the market, they are compelled to assume large sizes. Thus bigness happens very often to be a requisite for the viable operation of conventional economies. There may be a romantic flavor in the claim that 'small is beautiful.' In fact big is also commendable in its own right. Isonomies are typically social settings of moderate proportion having rigid intolerances for deviations of size beyond a certain range. Phenonomies are the smallest kind of social setting conceivable. A phenonomy may consist even of a single person, as is the case of the atelier of the painter or sculptor. But it seems doubtful whether a phenonomy may tend to keep its capability to endure when its members are more than five. Since anomy is not a social system, only indirectly does it pose questions of size; the treatment of anomic persons, however, is more successful in small social systems where they can be given personalized attention. Indeed the size of social systems in general has a bearing upon the scope of anomy in a given society. The size dimension of mass societies is itself a factor conducive to prompting anomy, since in it interpersonal relations tend to become predominantly functional, rather than affective. In modern industrial societies, as Émile Durkheim has highlighted, anomic types of conduct are necessary outcomes of the process of social division of labor. Yet one should not draw from Durkheim's findings support for an unqualified indictment of industrialization. The practice of social systems delimitation may very well be the corrective to anomy which has become a normal outcome of industrialization everywhere.

COGNITION

One pioneering work on the cognitive dimensions of social systems has been developed by the sociologist Georges Gurvitch (1955, 1971). He contends that there is a variety of types and forms of knowledge which rank themselves in a different priority sequence according to the nature of social systems. Thus archaic, feudal, capitalist, and socialist societies can be differentiated according to their specific dominant cognitive systems, i.e., the order of decreasing or increasing priority of types and forms of knowledge prevailing in each of them.

Although I find Gurvitch's comprehensive typologies quite useful for macrosocial analysis, I shall use a simpler and ad hoc procedure which I believe is significant to characterize briefly cognitive dimensions of the settings portrayed by the para–economic paradigm.

Habermas reappropriated the idea that cognitive systems can be classified according to their dominant interests.[6] For the purpose of this chapter it is enough to point out that a cognitive system is essentially functional when its dominant interest is production or the control of the environment; it is essentially political when its dominant interest is to enhance the good of the social whole; it is essentially personalogic when its dominant interest is the development of personal knowledge. A diffracted cognitive system is one deprived of a single central interest.

In a variety of mixes these cognitive systems may exist simultaneously in a single social setting. However, the functional cognitive system is predominant in economies, the political cognitive system is predominant in isonomies, the personalogic cognitive system is predominant in phenonomies, and finally the diffracted cognitive system is rather characteristic of anomic individuals and/or groups. Concretely, there are social systems in which more than one kind of cognitive system assume parallel dominance. Such is the case, for instance, with economies of isonomic character and many educational institutions where personal knowledge and the enhancement of the societal good are of paramount importance.

The main inference to be drawn from these tentative statements is that the all-inclusiveness of the market system in a society such as ours, continuously involving individuals in its inherent cognitive patterns, may disable them from acting as effective members of phenonomies and isonomies. Accordingly, in designing these systems and their mixes one should strive to provide individuals with conditions fit to their specific dominant cognitive interest.

SPACE

In its expansion throughout the last two centuries the market system has increasingly pre-empted the spaces reserved to social systems constituting the thrust of personal and community life. The architecture of contemporary cities meets market demands *par excellence*. One might recall that in its early stage the market system induced in England the practice of enclosures which evicted a great mass of people from their customary spaces. The industrial revolution has coerced populations to change from roomy households and cottages to narrow flats and apartments, and overcrowded buildings and ghettos near the urban centers. In this process, people lost time, money, and their direct relationship with concrete natural contexts.[7] In other words, deterioration of the conditions of people's community life has been a normal outcome of the expansion of the market system. The small family is a symbol of this transformation. It implies that contact between the old and the new generations is largely discontinued. Grandparents and grandchildren, thanks to the shrinkage of space, are supposed to live apart, a fact which in itself has undermining effects upon community life. It is no wonder that today people very often express a certain nostalgia for the old times. As de Grazia points out 'the ... space Englishmen had at the coming of industry [is] lost. Englishmen and Americans pay up for it everyday. Yet wherever a park is opened up it is blushingly heralded as a triumph of good government or philanthropy, and progress in any case. The partial recovery of lost ground becomes progress' (de Grazia 1964: 328–9). Presently, recovery of space for personal and community life should be a priority goal of citizens and governments, demanding a qualified delimitation of the market system.

Space has been closely scrutinized by organizational experts usually as a dimension of the process of production and delivery of goods and services. But it has manifold implications extending far beyond such economic intents which very often escape the perception of the layman and even of those, like architects and organizational experts, who supposedly should thoroughly understand them. Space affects, and to some extent even shapes, people's lives.

No wonder that individuals of great sensitivity to conditions which affected their personal unique story very often highlight among them the houses in which they lived. For instance, in his autobiography Goethe portrays the house where he lived as a child as having a benign influence on his process of growing up. He acknowledges that the house awakened in him 'a feeling of solitude, and resulting from it a vague yearning which corresponded to [his] temperament' (Goethe 1949: 4). Hermann Hesse refers to a house in which he lived for twelve years, which he names the Casa Camuzzi, as a place where he 'enjoyed the deepest solitude and suffered from it' (Hesse 1973: 246) and could give free rein to his imagination and creativity. The sensitivity to space is certainly one of the reasons which made Axel Munthe's *The Story of San Michele* a fascinating autobiographical account. When he decided to build his ideal place his awareness of the design it prescribed was so imposing that he dared to defy the advice of the constructor, Maestro Nicola. He reports: 'I told Maestro Nicola that the proper way to build one's house was to knock everything down nevermind how many times and begin again until your eye told you that everything was right. The eye knew much more about architecture than did the books. The eye was infallible, as long as you relied on your eye and not on the eye of other people' (Munthe 1956: 436). No doubt this precept of space design works effectively for individuals like Axel Munthe, who are extremely clear about their existential agenda.

It is hard to imagine a more vivid expression of a spatial feeling than the one which is offered by Carl Jung in a chapter ('The Tower') of his *Memories, Dreams Reflections*. Like Axel Munthe, late in his life Jung decided to 'put his fantasies and contents of the unconscious on a solid footing' (Jung, 1963: 223). He materialized his wish on a piece of land bought in 1922 in Bollinger (Zurich). Jung speaks of the house he built on this place, in several steps, as his 'confession of faith in stone' (223), the representative of his 'maternal hearth' (224). One does not need, however, to be so resourceful and abundantly gifted as these men were to realize that spaces where we happen to live may nurture or hinder our sound psychic growth as unique persons.

Space may be a factor facilitating or inhibiting the release of tensions as well as a determinant of stress. Animal behavior studies have indicated that each animal species seems to need an adequate space in order to carry out normally the activities inherent in its specific type of life; exposed to conditions overtly in violation of a required adequate space, any animal develops perverted patterns of conduct. One of the most impressive demonstrations of this fact is the massive death rate among rats living in crowded space observed among wild Norwegian rats even when available food was plentiful. There exist abundant but scattered observations of the effects of space on human life, although only recently have efforts at their systematic study been undertaken.[8]

I shall now try to indicate some implications of the space aspect for social settings design. I argue that *specific requirements of space dimensions are inherent in each type of social setting*. Thus the spacing of social systems is an art as well as a science. Its practice, formally or intuitively, demands what Fred I. Steele calls environmental competence, i.e., 'a person's ability to be aware of the surrounding environment and its impact on him; and ... his ability to use or change his setting to help him achieve his goals without inappropriately destroying the setting or reducing his sense of effectiveness or that of people around him' (Steele 1973: 113). Steele conceives environmental competence as a trained capacity since it has specific operational correlates which can be systematically learned.

In his books *The Silent Language* and *The Hidden Dimension* Edward T. Hall has focused upon relevant aspects of the human lifespace from the anthropological viewpoint. He calls attention to H. Osmond's distinction between *sociofugal* and *sociopetal* spaces (Hall 1966: 101), i.e., those which keep people apart, and those which facilitate and encourage conviviality. Neither kind of space is intrinsically good or bad. They are needed for different reasons. Hall states: 'What is desirable is flexibility and congruence between design and function so that there is a variety of spaces, and people can be involved or not, as the occasion and mood demand' (103–4). What should be avoided is the inadvertent aggravation of centrifugal dimensions of space in social systems where they should be centripetal. Hall points out: 'Virtually everything about American cities today is sociofugal and drives men apart, alienating them from each other. The ... shocking instances in which people have been beaten and even murdered while their 'neighbors' looked on without even picking up a phone indicates how far this trend toward alienation has progressed' (163). The predominace of sociofugal spaces in American cities like, for instance, Los Angeles, New York, and Boston, qualifies them as true 'behavioral sinks,' an expression which Hall borrows from John Calhoun (24). Living in these cities demands from individuals a great amount of psychic energy to countervail the pressures fostering pathological behavior.

Architects and designers have long stressed that space may be a factor of human deformation. It is said, for instance, that Frank Lloyd Wright claimed that 'he could design a house in which after six months any marriage would collapse' (Skolimowski 1976: 84).[9] In a paper where

Henryk Skolimowski analyzes the urban designs undertaken by Paolo Soleri in Arizona he submits, though not in reference to Soleri, that 'many architects are designing just [this] kind of house without knowing Frank L. Wright's formula' (Skolimowski 1976: 84). Skolimowski describes Arcosanti, an experimental commune in Central Arizona designed by Soleri, as a living setting where human nature and environment are adequately matched. The flaws of the American urban scenario have been thoroughly pinpointed by many social critics, among whom Paul and Percival Goodman deserve special attention. In their book *Communitas* they go beyond social criticism and articulate models of urban design which embody sound principles bearing upon the art of matching space and human needs.

In the works of space analysts like Hall (1959, 1966), Sommer (1969, 1972), and Steele (1973), there is a wealth of suggestions which designers of para-economic social systems may find constructive. Particularly relevant to this effect is the classification of physical features which both Hall and Steele conceptualize. Designers should learn to utilize fixed-feature-space, semi-fixed-feature-space, and pseudo-fixed-feature-space, i.e., learn when and where a physical setting requires features that are immovable or relatively immovable (load-bearing walls, monuments, buildings, streets, floors), flexible and movable features like chairs, pictures, desks, rugs, and drapes; and apparently immovable features. There is in Steel's book *Physical Settings and Organization Development* a store of practical instructions on the task of spacing social systems.

Designers of social systems like isonomies and phenonomies and their possible mixes should realize that adequate spacing is an essential condition for their successful functioning. Space speaks a silent but eloquent language by which people are inadvertently affected. It is said that in a debate about repairing the damages to the House of Commons building caused by the war, Churchill expressed the desire of preserving the traditional House site, where representatives could not avoid facing each other when speaking, lest its redesign would alter the patterns of government. 'We shape our buildings,' he said, 'and they shape us' (Hall 1966: 100). A lady pointed out to her husband: 'If any of the men who designed this kitchen ever worked in it, they wouldn't have done it' (98). Thus even the sex of the designers may unknowingly influence the spacing of social systems. It may very well be the case that one of the main architectural outcomes of the women's liberation movement will be the reinterpretation of the functions of the kitchen. It now has become less a place for cooking than a site for intense and intimate social relations among people, irrespective of sex and age. Moreover, women and men share its operation on increasingly equal terms.

Finally, to the extent that one of the fundamental objectives of social systems delimitation is to contain the influence of the market system on human life-space, its practitioners, more than anyone else, must be aware of the law of requisite adequacy. It seems evident that the market system conditions American citizens' perception and use of space. For instance, as noted by Hall, 'Americans pay attention to direction in a technical sense, but formally and informally, they have no preferences. Since our space is largely laid out by technical people, houses, towns, and main arteries are usually oriented according to one of the points of the compass ... A technical pattern which may have grown out of an informal base is that of positional value in almost every aspect of our lives, so much so that even children four years old are fully aware of its implications and are apt to fight each other as to whom will be the first' (158–9). In his study on material affluence and its bearing upon the formation of the American character, *People of Plenty*, David M. Potter has noted that 'the household space provided by the economy of abundance has been used to emphasize the separateness, the apartness, if not the isolation, of the American child' (Potter 1954: 197). In the United States today, when optimization and conservation of resources have become a public concern and an item of governmental agenda, the influence of our culture on individual perception and use of space should be brought under systematic attention by policy makers and designers. Americans should learn to transcend the market's proclivity to exploit the use of space if they are seriously committed to overcoming the ecological deterioration of their society. The affective ties of the individual to his physical environment, as demonstrated by Yi-Fu Tuan in his book *Topophilia* can be the object of scientific inquiry. They can be identified and categorized in each culture, as well as appreciated as a variable bearing upon the ecological effectiveness of life-space designs.

Appropriate spacing of social systems is certainly one of the means to nurture the psychological atmosphere commensurate with the systems' specific objectives. Topics like solitude, privacy, reserve, intimacy, anonymity, personal territory, orbit, and others, are issues to be taken into account in spacing social systems, particularly isonomies and phenonomies. The art of designing these settings has much to gain if it incorporates the contributions which anthropologists and environment psychologists have to offer.[10] Apparently, sociopetal rather than sociofugal spaces should prevail in isonomies and phenonomies as well as in settings designed to resocialize anomic individuals. Because of the nature of their activities, economies are systems in which sociofugal spaces are expected to prevail, although in a limited scope sociopetal spaces also are functionally required in such settings.

TIME

I now turn to focus upon time as a dimension of social systems. I must point out, the fact that I treat time and space in separate sections does not reflect my acceptance of the Newtonian split of space and time. Only because of an imperative of ordered exposition does one theme follow the other. Space and time are mutually implicated. The temporal orientation of the members of a social system has intrinsic space correlates. The spacing of social systems, on the other hand, induces specific temporal orientations.

Time as a category of organizational design has been a theme of standard organizational scholarship. However, in such a domain only the time inherent in economizing systems has been the object of study. Thus Taylor and some of his associates were pioneers of the study of time and motion as an aspect of scientific management. But the time they focused upon is a limited case constituting one aspect of the temporal spectrum of human experience. In this tradition most studies of time now available in the domain of organizational scholarship do not transcend the Taylorian conception.[11] They deal only with time as a commodity or an aspect of the linearity of organizational behavior. Important as this facet of the human experience of time is, it is not the main thrust of a variety of social systems like isonomies, phenonomies, and their mixes with economies. Thus the para-economic paradigm prescribes a multidimensional approach to time as a category of social systems design.

In the domain of sociology Georges Gurvitch (1964) has developed a typology of time dimensions of social systems.[12] He contends, for instance, that the time of formal organizations is not identical with the time characteristics of social systems where intimacy and intense interpersonal reciprocity prevail. Furthermore he conceptualizes a variety of time to explain the differences of human orientation in the feudal, capitalist, and social societies. The time dimensions of the social system from the para-economic viewpoint can only be presented tentatively. A typology can be proposed as constituted by the following categories: serial, linear, or sequential time; convivial time; leap time; erratic time.

Economies are settings in which serial time prevails, and thus they are unable to meet the human needs whose satisfaction involves a time experience which cannot be serialized. The market-centered society tends to serialize its members' time according to its temporal orientation, and largely succeeds in doing so, thus developing in them a trained incapacity to engage in endeavors requiring other kinds of temporal orientation. 'Americans,' says Hall, 'think it is natural to quantify time. To fail to do so is unthinkable. The American specifies how much time it requires to do everything' (Hall 1959: 134). Defining monochronism as the tendency of 'doing one thing at a time,' Hall states that 'American culture is characteristically monochronic' (138). He compares this American cultural trait with the polychronism of other cultures:

In the Silent Language, I describe two contrasting ways of handling time, monochronic and polychronic. Monochronic is characteristic of low-involvement peoples, who compartmentalize time; they schedule one thing at a time and become disoriented if they have to deal with too many things at once. Polychronic people, possibly because they are so much involved with each other, tend to keep several operations going at once, like jugglers. Therefore, the monochronic person often finds it easier to function if he can separate activities in space, whereas the polychronic person tends to collect activities. If, however, these two types are interacting with each other, much of the difficulty they experience can be overcome by the proper structuring of space. (162)

To be sure, the Western assessment of the time orientation of people living in peripheral and primitive societies as an indicator of laziness or lack of achievement motivation is nothing but an expression of cultural parochialism.

Membership in social settings other than economies requires psychological propensities which very often many individuals fail to develop. An all too vivid example of this failure is the mass of retired people in American society who do not know what to do with themselves when they lose the condition of jobholder. 'Americans are confused about work,' says A.K. Bierman. He adds, 'Unless we can learn to go to bed with the machine in Eden, it will be our dehumanization rather than our benefactor' (Bierman 1973: 15). It may be that the dominant temporal orientation of most Americans is the main factor which hinders them from engaging in isonomic acculturation processes.

Isonomy is a site for conviviality, and its paramount temporal requisite is an experience of time in which what individuals gain in their relationships with others is not quantitatively measured. Instead a deep gratification is felt for being liberated from pressures which preclude their personal actualization. Convivial time is cathartic. The individual's experience of convivial time encourages him to interact with others without facades, and vice versa. When a group of people share this kind of temporal experience they ease up, tend to trust each other, and authentically express their deep feelings. The participants of this social interaction do not see and treat each other as objects, but as persons. They accept and appreciate each other for what they are, irrespective of their corporate roles or status in the competitive environment of the market. One forgets time in the serial sense when one engages in the experience of convivial time.

Leap time is a very personal kind of time whose quality and rhythm reflect the intensity of the individual's urge toward creativity and self-enlightenment. It is an eventful momentum in the life of a creative and searching person, alone or together with other persons who are also attuned to the same sort of inquiry. It is the temporal thrust of phenonomies.

Leap time does not belong in the realm of *chronos*. The Greek mind conceived *chronos* as a dimension of the temporally restricted and regulated part of the cosmos, beyond which there is what Anaximander called the *apeiron*, i.e., the boundless, the limitless, where all things ultimately come from.[13] It seems to me that it is from the latter that leap time emerges to become part of the realm of *kairos*, a Greek word which denotes a time which is not quantifiable and is constitutive of human insights into the process leading to critical happenings. *Kairos* is a feature of a certain kind of person's inward life when involved in self-exploring journeys and/or endeavors culminated by breakthroughs. Psychologists acknowledge this kind of time as a *dactum* of human experience in certain circumstances. It has some semblance with what Laing (1967) calls *eonic time*, a characteristic of deep subjective events (Laing 1967: 128). It is also a central concern of Jung and Progoff in their approach to the human psyche.

Jung tells of events of his very life as occurring 'outside time' (Jung 1963:225) and belonging to the 'spaceless kingdom' (226) of the psyche. He suggests that the meaning of such events is apprehended in the context of symbolic experiences in which, as Progoff explains, 'timeless images' and 'spaceless space' prevail (Progoff 1973: 53, 135). When engaged in symbolic experiences the individual reaches out of the immediate social confines of everyday life. It is in this sense that we must understand that all socialization is alienation from the inward world of the psyche. Socialization has contradictory facets: without it, the individual does not survive as a member of the species, but when entirely captured by it, he or she loses the character of a person.

Both Søren Kierkegaard (1962) and Henri Bergson (1956) have described a type of human creative experience which takes place only when the individual succeeds in breaking the confines of the social. The experience involves a *leap* from the closed into the open (Bergson 1956: 77), from the stifling norms characteristic of a peculiar age into eternity. For Kierkegaard in particular the leap is tantamount to the person's self-discovery. 'By leaping into the depths,' he says 'one learns to help oneself' (Kierkegaard 1962: 58). Obviously, because of its transsocial character the content of existential leaps can only be articulated through symbolic language. One might argue that the domain of symbolic experience is extraneous to organizational scholarship. I contend, however, that any organizational scholarship oblivious to symbolic experience fails to exercise its humanistic role. Truly humanistic organizational scholarship must be critically aware that social models of man are always categories of expedience. But expedience is not the only concern of organizational scholarship; it must be sensitive to what in the human being is irreducible to sociality, in order to enable him to resist the fluidity of his psyche and his deformation as purely a specimen of episodical corporate life. This scholarship must be able to help the individual keep a sound balance between the external requirements of his corporate condition and his inward life. Thus serial time must be recognized for what it is, and not mistaken for all that time is.

There is an abundance of scholarly sources about the experience of leap time. In men's and women's self-motivated endeavors, the incidence of leap time is marked by ups and downs of the individuals' state of spirit, and is experienced as a mix of suffering and joy. The downs may be deeply depressive, but they represent the necessary steps which individuals have to go through in order to consummate self-rewarding goals. When the sufferings which a successful person has undergone in a creative pursuit are over, they are looked at as gratifying experiences. After the demanding ordeal of a successful creative deed, people usually indicate that they would go through the same steps if they were put in the position of choosing again. Leap time is an eventful moment of self-gratifying creative pursuits.

Leap time frequently occurs in breakthroughs undertaken by creative people, including inventors, reformers, administrators, scientists, novelists, painters, and poets. A pattern can be described in their careers: (1) in general they are people who enjoy and know how to work with themselves in solitude (which phenonomies are meant to protect); (2) they seem to have a keen awareness of their calling; (3) they keep themselves busy as if moved by an inner compulsion (a fundamental indicator of leap time) which enables them to accomplish what is beyond the ken of ordinary persons.

Some expertise already exists for designing settings which seem to have the characteristics of phenonomies. Research and development teams working in business corporations and organizations like the RAND Corporation and NASA, and special kinds of governmental task forces, would not be successful if adequate skills for managing phenonomies were not available.[14] Yet, concentrated effort now needs to be organized to develop and consolidate an expertise for creating social systems of a more alternative character than those envisioned by the think tanks funded in institutions representative of the market system.

I call a time of inconsistent direction erratic time. Those affected by erratic time have a diffused experience of their existential agenda, if they have any agenda at all. Circumstances other than their will to meaning directly shape the course of their lives. To illustrate concrete cases of erratic time one can immediately think of anomic and quasi-anomic people such as beggars and the excluded who ordinarily find their milieu in skid row areas, migrant workers, hoboes, peddlers, and, in some cases, unemployed and retired citizens. Yet to live according to the vagaries of erratic time may temporarily be conducive to personal growth. It is presumed that much of what George Orwell narrates in *Down and Out in Paris and London* is autobiographical. And certainly his experience as a penniless writer in those cities during his youth taught him to understand better himself and his vocation. Knut Hamsun's *Hunger* seems to exhibit a similar type of youthful experience in Oslo. Of identical significance is Ernest Hemingway's *A Movable Feast*, the narration of his self-searching days as a young man living in Paris. Tourist and travel agencies seem to know how to organize journeys in order to refresh people looking to be temporarily exempted from the obligation of caring about what to do next. To be sure, a *know how* meant to recuperate into the mainstream of society citizens systematically caught within the syndrome of erratic time should be an issue for designers of alternative social systems.

One of the objectives of this typology is to lay bare the process of unidimensionalization of time which victimizes most persons living in the market society. Standard economic and organization theory focus upon time within a narrow unidimensional perspective. They deal only with serial time, systematically overlooking human pursuits not functionally prescribed by the market system. They accept the social time inherent in the market as determinative of the nature of societal temporality at large. It is precisely this situation that para-economic policies and designs seek to overcome. Individuals who are overaccommodated to the temporal orientation intrinsic in the market can hardly understand the extent and nature of their psychic deformation. A therapy meant to heal this deformation may possibly be developed as a set of procedures to help people engage in non-serial experiences of time. A completeness of human existence is forfeited if one does not address the substantive topics constitutive of one's predicament according to the varieties of time the predicament implies.

The synchronization[15] of human life with the requirements of the market system, dominant in contemporary industrial societies, is a chronic factor of a very well identified common pathology, i.e., high incidence of apathy, alcoholism, drug addiction, insomnia, nervous breakdown, stress, suicide, anxiety, high blood pressure, ulcers, and heart disease. One should try to understand the message of this normal pathology. The market-centered society has deprived the individual of the variety of experiences of time he used to find available. In the past people had plenty of opportunities to engage in genuine conviviality and personal endeavors, completely free from any relations with the market-place.

In providing its members with those opportunities, past societies interpreted themselves as replicas of the cosmos and as conforming to sacred or quasi-sacred prescriptions. In these societies people had plenty of time, unrelated to their condition as laborers, to engage in self-gratifying pursuits. In their calendar the character of hours, days, months, and years reflected their concerns with the multiple overtones of the sacred dimension of life. In the middle ages 167 days of each year were non-work days, including 52 Sundays. Days for fêtes and holidays exceeded working days in Greece in the late first century BC. At about the same period in Rome 65 days were reserved for games. In Rome by the second half of the second century AD games took 135 days and later in the fourth century 175 days.[16] Today one can hardly capture the sense of festivity and celebration that animated those calendar spots.[17] By contrast, it is a fundamental fact in contemporary societies that there is no day in the calendar which is not pervaded by the temporal prescriptions inherent in the market. The market has pre-empted the functions of sacral agencies and become the arbiter of temporality at large.

Such a synchronization should be reversed, gearing the market to function in consonance with the requirements of social systems which enhance the quality of community life in general, conviviality, and the citizens' personal actualization. This task has been undertaken by many citizens in this country involved in a multifaceted gamut of alternative social experiments. The study of the policy implications of this and related trends will be undertaken in the next chapter.

9

Para-economy: Paradigm and Multicentric Allocative Model

For several reasons the paradigm presented here is called para-economic. So far I have used this word to qualify an approach to social systems analysis and design in which economies are considered as only a part of the total social fabric. However, para-economy can also be understood as providing the framework for a substantive political theory of resource allocation and functional transactions between social enclaves which are necessary to enhance qualitatively the citizen's social life. A systematic statement of this theory has not yet been developed, although fragmentary contributions to its elaboration are already available. Moreover, in speech and deed there are many whose activities qualify them as para-economists, i.e., individuals who are attempting to implement scenarios which represent alternatives to market-centered processes.[1]

In opposition to the prevailing market-centered approach to social systems analysis and design, the para-economic paradigm postulates a society diversified enough to allow its members to deal with substantive life issues according to their pertinent intrinsic criteria and in the specific settings where those issues belong. From the para-economic political viewpoint, not only economies which already constitute the market enclave, but also isonomies and phenonomies and the variety of their mixed forms are to be considered agencies through which manpower and resource allocation is to occur. It is in this latter sense that social systems delimitation is applicable at a societal as well as at a micro-organizational level. In other words, like economies, isonomies and phenonomies must also be considered legitimate agencies required for the viability of society at large.

There are two basic ways to implement societal allocative policies and decisions: two-way transfers which characterize the exchange economy and one-way transfers which characterize what Kenneth Boulding and his associates call the grants economy.[2] For example, there are social systems, mainly those employing an exchange allocative mechanism to deliver standard goods and services to the public, whose effectiveness is evaluated through standard price-profit accounting. Enterprises like factories, shops, or professional offices, can only keep themselves in business if they find clients willing to buy their products or pay for their services, and thus provide them with the margin of profit necessary for their ongoing operation. But the quality and the growth of a society does not result only from the activities of such market-centered systems. Quality and growth also result from a variety of outputs delivered by non-exchange allocative processes. Examples of these outputs are those delivered by households, neighborhood associations, churches, pioneering altruistic initiatives, regional and national campaigns to meet such neglected needs as protection of women, children, youth, and senior people, and networks of concerned citizens focusing upon human rights, the environment, and other public issues. The evaluation of the effectiveness of these undertakings involves more than a direct accounting of factors of production. Their contribution to the viability of the social whole is unassessable in a conventional cost-benefit framework. They normally cannot function unless funded by grants. The complex questions of what activities in a society should be funded by grants or organized by exchange criteria, and what kind of political support a state needs to accomplish delimitative functions is beyond the scope of this analysis.[3]

Prevailing allocative models are predicted upon a very narrow conception of resources and production, which are understood as only inputs and outputs of activities of an economizing nature. In other words, it is the market which in the last analysis dictates what is to be considered as resources and production. Accordingly one does not formally consider as contributors to the national wealth the undertakings of household members who, without earning a salary, engage in activities like cooking, cleaning, sewing, growing vegetables, making preserves, gardening, decorating, caring for the sick, repairing and fixing, and child education and supervision. Likewise the citizen who, without being paid, participates in neighborhood church meetings, artistic ensembles, educational encounters, and self-help endeavors of all sorts is not accountable as a resource. In peripheral countries a great part of the population who labor as 'peasants' are conventionally not considered to be productive as long as the output of their activities is not commercialized. Nevertheless household members, the concerned citizen, and the self-providing peasants in fact do *produce*. For instance, it has been estimated that the value of household labor in the U.S. represents approximately a third of the Gross National Product and half of disposable consumer income (Burns 1975:14). Yet because the output of household labor is not directly transferable to the market it is ignored by the official system of statistics which implies that production is equivalent to selling and that consumption is tantamount to buying. In computing the 'wealth of the nation,' this system registers only what is sold or bought. No wonder that an immense reservoir of resources and productive capacity is overlooked and left untapped by prevailing allocative models.

By comparison, the para-economic paradigm implies a conception of production and consumption which formally accounts for both remunerated and non-remunerated activities. The productive individual is not necessarily a jobholder. The identification of the one with the other constitutes one of the main fallacies and blindspots of prevailing allocative models. Another fallacy and blindspot is the assumption that the amount and quality of the citizens' consumption is expressed in what they buy. In fact the market largely ignores what people need and only 'knows what people can be made to buy' (de Grazia 1964: 215). In other words, the market only produces what it 'can sell' (215). Caught within these fallacies and blind spots, governmental allocative policies have been unable to reach out of the vicious circle of the market system in order to take advantage of existing possibilities of building a variety of cashless productive settings as part of a multicentric society.

There is today a widespread concern with the problem of 'finite resources.' Indeed it is true that a number of critical physical resources which the market system needs in order to continue to operate are non-renewable and may be exhausted in the long range. But the prevailing understanding of this problem is distorted, leading, for instance, to the concept of the 'limits to growth.' This is a misnomer. In fact, as suggested above, a sound concept of resources includes more than what the market is bound to define as a resource. It includes ecological and psychical dimensions to which the mechanistic epistemology inherent in the classical law of supply and demand is insensitive. In the same vein, the argument in favor of 'zero growth' strategies is largely an admission of the bankruptcy of the present configuration of the market system. I submit, however, that limits to current market activities are not necessarily limits to growth. There are plenty of resources and substantial productive capacity which stay idle because of a lack of an adequate theoretical scheme for organizing these potentialities. From the para-economic standpoint resources are infinite and there are no limits to growth. Ironically the 'limits to growth' thesis may very well represent the opportunity for revealing a vast horizon of possibilities for a growth explosion in terms of both production and consumption. In order to realize these possibilities, individuals, institutions, and governments have to rid themselves of the conceptual blinders inherent in the market-centered allocative models. In general the main assumptions of these models can be articulated as follows:

1 / Criteria for assessing the development of a nation are essentially the same as criteria for the activities constituting the dynamics of the market. Accordingly, the size of the GNP as conventionally conceptualized, the percentage of citizens living in urban areas, and the percentage of the labor force employed in the service sector are all taken as the relevant indicators of development.

2 / Human nature is assumed to be defined as the set of qualifications and dispositions characteristic of the individual as a jobholder and an insatiable buyer. Thus the socialization process in particular must be geared toward developing the citizens' potential to succeed as jobholders and their capacity to prove their worth through the comparative rank of their purchases.

3 / The effectiveness of organizations and institutions in general is assessed from the standpoint of their direct or indirect contribution to the maximization of market activities. This approach leads to unidimensional types of organizational theory and practice, and policy science models of which conventional 'public choice' and current economic theories are illustrations.

Dissatisfaction with these models has been voiced in many corners of the academic world. It is worth pointing out that 'policy science' and 'policy analysis,' as the expressions are conventionally understood, largely consist of an attempt to focus upon the process of policy formulation, implementation, and evaluation from the standpoint of the instrumental rationality inherent in the classic economic calculus. It is no wonder that authors who subscribe to those policy models strive towards enhancing the 'scientific' character of political theory by borrowing concepts from the fields of operational research, systems analysis, cost-benefit analysis, and computer technology, and by assuming that quantitative approaches and methods are the best, if not the only tools for improving the study of policy making.[4] The limited character of this orientation has been successfully highlighted by several scholars.[5] Nevertheless this critical effort has not yet resulted in an alternative to the conventional policy science model. I submit that the para-economic paradigm is, at least, an incipient broad theoretical framework for such an alternative.

The para-economic paradigm adds two essential qualifications to the policy science/policy analysis discussion. First, it assumes that quantitative methods have the highest probability of being useful in the study of ecologically sound profit-maximizing and/or satisfying policies; however, these policies in turn are seen as a restricted area of concern in the domain of policy science. Second, the para-economic paradigm asserts that there are normative and substantive allocative policies which are indispensable if we wish to enhance the qualitative status of the macrosocietal system. In other words, the utilization of conventional policy models must be consistent with the law of requisite adequacy. We must recognize that these models assume an ideological overtone when they step beyond the specific context of the market enclave and aim at subsuming under their criteria the entire social process of resource allocation.

For illustrative purposes it is helpful to reappraise the work of Robert Dahl and Charles Lindblom. Their elegant and significant book *Politics, Economics and Welfare*, published in 1953, has more in it than the authors themselves have explored. They conceptualized four models of choice and allocation: (1) the price system (control of and by leaders) (2) hierarchy (control by leaders), (3) polyarchy (control of leaders), and (4) bargaining (control among leaders). My contention is that disentangled from their overarching economizing intentionality these models could very well provide the theoretical underpinnings of the allocative process and serve as tools of a multicentric policy system. In particular, hierarchy and polyarchy could be envisioned as allocative models categorizing governmental functions required to nurture isonomic and phenonomic enclaves and protect them against the distortive penetration of the market system. A careful reading of their book suggests that the authors show a high sensitivity to substantive issues of resource allocation. Yet, because economizing is the overarching concern of their book, the para-economic character of their models fails to be clearly spelled out. Had the authors developed a systematic distinction between substantive and functional rationality and their policy implications, it is likely that they would have come close to articulating much of what constitutes the para-economic paradigm.

For instance, Dahl and Lindblom use the expression 'rational action' and 'efficient action' as interchangeable, i.e., as 'designed to maximize goal satisfaction' to the extent that 'goal satisfaction exceeds goal cost' (Dahl and Lindblom 1963:39). At the same time they would wish the reader to understand that there are 'goal costs' and 'goal satisfactions' like leisure and conviviality which cannot be measured by 'quantitative symbols as "efficient"' (40). Dahl and Lindblom indirectly acknowledge the reality of the isonomic and phenonomic enclaves when they point out that 'it is on small groups that most people rely for love, affection, friendship, "the sense of beauty" and respect,' and note that these groups 'carry on the main burden of indoctrination and habituation in identifications and norms, transmitting the habits and attitudes appropriate to polyarchy' (520). More specifically, they state:

In so far as it is attainable at all, for most people much of 'the good life' is found in small groups. Family life, the rearing of children, love, friendship, respect, kindness, pity, neighborliness, charity: those are hardly possible except in small groups. *If one could somehow destroy the large groups and leave these things standing, the loss of the large would be quite bearable* [emphasis mine – A.G.R.]. But if one maintained the large groups and destroyed these values, the impoverishment and barrenness of living would be incalculable. (520)

The Dahl and Lindblom approach to resource allocation is largely correct as long as it remains within the qualified boundaries of the economizing enclave. However, as it stands, it is an arrested statement of choice and resource allocation because in it isonomies and phenomonies, under the expression of 'small groups,' are only incidentally, and not systematically acknowledged as categories for ordering the process of policy making.[6]

In contradiction to market-centered models, the para-economic paradigm provides a systematic framework for developing a multidimensional and delimitative thrust to the policy-making process. This paradigm, focusing upon resource and manpower allocations in macro- and microsocial systems, assumes that:

1 / The market should be regulated politically and delimited as an enclave among other enclaves constituting the total social fabric. In other words the market has inherent criteria which are not the same for other enclaves and society at large. Further, the quality of social life in a nation results from the productive activities which enhance its citizens' sense of community. Accordingly these activities are not necessarily to be assessed from the standpoint inherent in the market. Thus social systems delimitation leads to strategies for allocating resource and manpower at

the national level which reflect a functional integration of one-way and two-way transfers. A public policy formulation, economic planning, and budgeting expertise congenial to social systems delimitation needs to be developed.

2 / Man's nature actualizes itself through a manifold of activities, among which are those required by his incidental condition of being a jobholder. Man's actualization may be inversely proportional to his consumption of market outputs and commodities, and most particularly to the time required for such a type of consumption. This conception of man implies that a completely socialized individual is necessarily less than what a person should and can be. It also implies that the educational system should be concerned essentially with the growth of individuals as persons, and only secondarily as jobholders. Moreover, insofar as unlimited consumption of market outputs is pollutive and depletive of natural resources, in the final analysis it must be considered as unethical.

3 / The development of effective organizations and institutions is assessed in general from the standpoint of their direct or indirect contribution to the strengthening of the individual's sense of community. This approach leads to a multidimensional type of political and organizational theory and practice which is conceptually and operationally qualified to enhance both the citizens' productive activities and their sense of meaningful personal and social actualization.

It is evident that nowadays there is in the academic milieux a widespread uneasiness with standard approaches to development.[7] These standard approaches are misleading precisely because they allow the market to be the cardinal referent for the process of resource allocation. Thus, for instance, they imply that an increase in the volume of exchange activities and a spatial expansion of the market are tantamount to development. This bias is particularly clear in the standard assessment of the phenomenon of 'dual economy' in peripheral countries. Thus it is said that a country where there is a 'dual economy,' or there are populations living in areas not included in the market, is, by definition, underdeveloped or even backward. The advice that policy makers in these countries usually receive from Western experts is that since the 'dual economy' constitutes a drawback to development, efforts should be undertaken to incorporate the whole population of a country into the market system. The overall result of this policy orientation, not only in peripheral, but in centric nations as well, is well known. Some consequences are misurbanization or overconcentration of population in great cities, increase in the rate of anomy, the aggravation of the behavioral syndrome with all its distortive psychological connotations, the dilution of the citizens' cultural identity, and the destruction of their craft competence which enabled them to autonomously guarantee their meaningful livelihood. Moreover, the economizing quantitative overtone of such a policy orientation leads their subscribers to legitimate the primacy of increasing the GNP over social equity and income distribution.

The conventional interpretation of the phenomenon of 'dual economy' is extremely short-sighted. Currently the phenomenon is understood as the coexistence in a nation of self-providing rural settings and profit-oriented systems. However, this type of dichotomy is a particular form of economic duality which is a normal feature of all contemporary nations. Indeed in all of them, including the United States, there are two kinds of productive systems, namely, the profit-oriented and the mutuality-oriented systems. Moreover, they are not always antagonistically related. To think so is to indulge in a very myopic reading of the phenomenon. One might consider, for instance, the United States. From the para-economic viewpoint the governmental economic policy makers in this country largely fail to actualize fully the potential of its productive structure because of their captivation by the market mind-set. Although neglected by policy makers, mutuality-oriented productive systems are a paramount part of the American economic structure. At present the mutuality sector is alive and growing through a mushrooming number of private initiatives.[8] The activities of this sector constitute the main focus of attention of journals like *Co-Evolution* and *The Futurist* and the writings of many persons, including Hazel Henderson, Scott Burns, J. Gershuny, L.S. Stravianos, D.L. Meadows, A.K. Bierman, and Marilyn Ferguson. The mutuality sector harbors much of the creative energy this country needs in order to overcome the stage of diminishing returns in which the market economy now finds itself because of ecological constraints bearing upon it. Governmental policy makers do not seem to realize sufficiently that American society is generating imaginative schemes of resource allocation which, were they bolstered by adequate systematic policies, would represent antidotes for the flaws of the economy in its present dystrophic state. Like the physician who treats a patient with a medicine which aggravates his disease, these policy makers try to correct, with traditional market correctives, the distortions of social life caused by the market system, like high rates of unemployment of a structural nature and inflation largely resulting from ecological determinants. They ignore the society's self-healing energies stored in the mutuality-oriented productive sector.

The obstructive character of the American governmental policy system is also reflected in its neglect of what Kenneth Boulding calls the grants economy. As he points out grants are now a substantial part of the funds available to finance the productive activities of the nation. He estimates that from 20 to almost 50 percent of production in America is funded by grants rather than by exchange (Boulding 1973:1–2). One should not be struck by the apparent vagueness of the estimate. Grants assume a multiplicity of forms, some of them very elusive, and thus their precise statistical accounting will never be possible. For example, there are grants of a very visible nature, like those provided by private and public foundations and many types of donors. They are the less difficult to account for and possibly represent the lower limit of Boulding's estimate. The higher limit of the estimate plausibly refers to a variety of 'implicit' grants, i.e., 'redistributions of income and wealth that take place as a result of structural changes or manipulations of prices and wages, licenses, prohibitions, opportunity or access' (49); and to the multifaceted gamut of grants which activate the mutuality-oriented productive systems, which eventually include isonomies, phenonomies, and their mixed forms.

There exists in the United States a dual economy of sorts which is constituted by the grants sector and the exchange sector. This duality is not an abnormalcy, and one sector should not be subsumed under the goal imperatives of the other. Both sectors should be envisioned in their distinct specific nature as accomplishing complementary and socially integrative functions. However, grants are largely mismanaged by private and public grantors captive to the market mind-set, and are adequately utilized mainly through trial and error pursuits carried out by concerned citizens. For instance, for structural reasons the exchange sector of the American economy is becoming incapable of producing conventional job opportunities in sufficient number to absorb the available labor force.[9] It is largely as a reaction against this trend that one should interpret the circumstance that during fiscal year 1977–8, 10 percent of the increase of jobs were held by self-employed citizens engaged in small-scale ventures, and that 50 million Americans now are members of enterprises of a cooperative nature.[10] I submit that the failure of the dominant market system to absorb fully the population of individuals at an active age is incorrectly interpreted by conventional private and public policy makers as a temporary vicissitude of the economy. One consequence of this interpretation is that people who are forced to join the mass of social welfare and social security recipients are socially downgraded because of their unemployed status, as if this condition necessarily reflected their personal incapacity. Existing concrete avenues leading to full employment of the labor force escape the attention of conventional policy makers precisely because of their captivation by the market mindset. An act of imagination could allow the reservoir of idle capacity represented by people without formal jobs to be mobilized into the mainstream of the American productive system through the allocation of the funds of social welfare and social security, not as mere benevolent help, but as grants to finance citizens' activities and creativity in socially recognized mutuality- and community-oriented ventures.

One objective of para-economic policies is a balanced allocation of resources. For instance, from the para-economic viewpoint, the existence of 'dual economy' in a country may eventually be an asset rather than a drawback. This is not to say that efforts should not be undertaken to develop the market in a given country. But the para-economic paradigm implies that the development of the market should be regulated politically so that it does not undermine the basis of isonomic and phenonomic enclaves. Moreover, this paradigm acknowledges that the overall economic improvement of a nation is compatible with what is considered a 'dual economy' condition, i.e., the coexistence of mutuality-oriented systems, where members produce for themselves a great part of goods and services they directly consume, and profit-oriented systems where members are essentially jobholders who draw from their salaries the acquisitive power to provide for themselves all the goods and services they need. The mutuality-oriented systems and the exchange sector are therefore not reciprocally exclusive. Both are to be systematically and simultaneously nurtured through an effective utilization of one-way and two-way transfers for the good of the society at large. One implication of this observation is that in peripheral countries rural living conditions must be considered in their own terms and protected against the unqualified and disruptive penetration of the market if one is to enhance their self-reliant capabilities. In summary, the general welfare of individuals living in a dual system can only be improved through a balanced allocation of resources, both as one-way and two-way transfers.

The welfare of citizens is a cultural category peculiar to each nation and is not measured by criteria common to all nations. Being a systematization of the thinking patterns inherent in the market system, conventional economics implies that criteria for assessing social welfare are the same for all nations. Accordingly, we witness governmental authorities of peripheral nations formulating and implementing allocative policies which are expressions of the syndrome of relative deprivation and the demonstration effect. The mind-set of these authorities and that of the middle sector of these peripheral nations thus contribute significantly to a perverted allocative system.

It is in this sense that conventional economics is the ideological component of the classic industrial revolution. At best it succeeds as a conceptual tool to explain processes characteristic of the market-centered society. It does not, however, provide the conceptual referent to understand and deal with basic allocative issues common to all societies. Although it incorporates contributions of thinkers native to France and other European countries, in its dominant terms it is essentially an Anglo-Saxon ideology. Since its beginning it was bound to become the main thrust of the cognitive politics through which Western hegemonic industrial nations have induced the rest of the world to conform to their expansive propensity.

In the last two decades the pollutive and depletive outcomes of the practice of this Anglo-Saxon economic ideology have generated, among certain scholarly quarters, a critical reappraisal of classic economics, and attempts to build a science of resource allocation as an ecological discipline. So far the most elegant and penetrating statement reflecting this orientation can be found in the writings of Nicholas Georgescu-Roegen. Indeed the fallacious character of standard economics has become increasingly obvious as some externalities of its systematic application have sensitized scholars concerned with the deterioration of the environment and the exhaustion of the reserves of critical non-renewable sources of energy. Significant as the studies of these scholars must be considered, more than topical analysis of the ecological distortions resulting from the practice of classic economics would be necessary to reorient the process of resource allocation on a world-wide scale. In response to this need Georgescu-Roegen articulates the assumptional foundations of the new allocative science.

Briefly, Georgescu-Roegen points out that available and accessible low entropy matter-energy, which ultimately is the input of whatever man produces, constitutes a limited planetary dowry. Since matter-energy has an irreversible propensity to assume high entropic states, production of goods and services, for the sake of prolonging mankind's existence as a species, should not accelerate such a propensity. Accessible and available resources are of two kinds, namely renewable, i.e., those of a biological nature that can be reproduced within relatively short natural cycles, as well as the energy received from the sun and the kinetic energy of the wind and waterfalls; and non-renewable resources, such as oil, lead, tin, zinc, mercury, and other minerals whose reproduction, if possible, would require long ecological cycles, so as to make them practically unavailable within the time confines of mankind's existence. Production of goods and services should be carried out through the maximum use of renewable resources and a minimal sensible use of non-renewable resources. The scarcity of non-renewable resources is not of a temporary nature. To manage their utilization and allocation through market mechanisms, i.e., as if they were to be priced according to the classic law of supply and demand, is an illustration of the utilitarian/hedonistic rule, *après moi le déluge*. Indeed, any parcel of non-renewable resource used in the productive process will be gone forever. This fact tells something about the depletive character of contemporary macrosystems. In the last ten years half of the total amount of crude oil ever produced has been obtained; and in the last thirty years half of the total quantity of coal ever mined has been extracted. Irreplaceable as these and other materials are, their market prices are thus but fictions. If the utilization of these materials continues at current rates mankind will soon be deprived of their use.[11] Because of its prevailing patterns of production and consumption, today's world faces an unprecedented rate of absolute ecological scarcity, the exponential increase of which may hasten the thermodynamical breakdown of the planet which, as a matter of fact, is inevitable at a certain point in time.

The para-economic paradigm takes into consideration not only the thermodynamics of production but also its social and ecological externalities. Thus it is an alternative to the classic allocative models (whether derived from Smith or Marx) and also provides the overarching framework for a new science of organizations. No less than a world-wide organizational revolution is needed to overcome the physical deterioration of the planet and the conditions of human life.

The institutionalization of a multicentric society is now in process in multifarious and inchoate terms. It may be aborted or, on the other hand, it may gain momentum with our increasing awareness of the distortive externalities generated by the market-centered society. In any event the future will be shaped either through the mere passive coping of historical agents with circumstances, or through their creative exploration of unprecedented contemporary opportunities. Most likely, a new society will come about through both ways.

While nobody can claim to have the precise view of things to come, it is essential that we delimit the impingement of economizing organizations upon human existence at large if we are to capitalize on contemporary possibilities. Because economizing organizations are precisely those which most draw inputs from the limited budget of non-renewable resources, they should be redesigned severely with an ecological concern in mind. Such organizations should be circumscribed as an enclave within a multicentric society which provides many settings for self-rewarding pursuits involving the least consumption of low entropy inputs.

The industrial world we live in also began as an objective possibility.[12] It was shaped throughout an accumulative process of institutional innovations deliberately undertaken by many individuals. We may now be in a similar incipient stage of institutionalization from which an alternative to the market-centered society – the multicentric or reticular society – may emerge.

10

Overview and Prospects
of the New Science

In this book I have exposed from a theoretical perspective flaws of extant organization theory and set forth the framework for a new general science of social systems design. This analysis has also exposed prevalent shortcomings of contemporary social science of which current organization theory is a part. In the first part of this chapter I summarize key points in my critical assessment of conventional social science. In the second part I characterize the new science of organizations as being endurance centered.

CONVENTIONAL SOCIAL SCIENCE

The root of the fallacious character of conventional social science is the concept of rationality which pervades it. This book focuses upon a distinction between substantive and formal rationality, a distinction which has been proposed by a few great contemporary thinkers but never thoroughly explored by them as a referent for differentiating two types of social science. The distinction should not be considered as a didactic exercise. It poses an existential dilemma to whoever chooses to be a social scientist. To be sure, in general the choice of either one of the horns of the dilemma is usually not conscious, but made for individuals by means of their socialization in academic milieux which themselves operate within dominant Western institutional parameters. What theoretically ruins conventional social science is not its formal character; rather it is ignorance of its parametric character, i.e., of its penchant for relying on a world vision inherent in a precarious historical climate of opinion. Therefore it is bound to crumble when such a climate of opinion loses credibility. As distinguished from formal rationality, substantive rationality can first of all never be captured within a definitional statement. It is only through the unobstructed experience of reality and its articulation that substantive rationality can be understood. One cannot even hope to understand it through the mere acquisition of packaged information. Secondly, social science predicated upon substantive rationality transcends episodal climates of opinion. In particular historical periods it may be over-shadowed, but never destroyed. This is not to say that substantive social science exists as a definitive conceptual body which has been formulated once and for all times. On the contrary it is always in the making, each epoch adding to and expanding the millennial legacy of insights into human nature and human associated life. The critical analysis I have presented does not have a literal restorative intention. Rather it is a call for the appropriation of that legacy and its development in terms which would enable us to understand and master the process of contemporary history.

The market-centered society and the social character that it engenders are recent events in history. They were formed in the wake of an industrial revolution consummated in a few centric Western countries. From the vantage point of this book we now understand that this transformation cannot be considered as the only path those countries could have taken during the last three hundred years. Caught within the illusionary interpretation of this accomplished fact as the outcome of a necessary unfolding of history, conventional social science postulates the market-centered society and its resultant social character as the yardstick to evaluate mankind's past and present history. Thus, in spite of its value-free claims, contemporary social science is normative to the extent that in theory and practice it is nothing more than a body of social systems analysis and design criteria induced from a particular historical configuration. Confronted with ecological constraints upon production and consumption – constraints which require a delimitation of the market system – the ideological underpinnings of conventional social science become increasingly evident. The alternative model of social science outlined in this book is not anti-market. Moreover my criticism of contemporary market-centered society should not be interpreted as an advocacy of the elimination of the market as a functional social system. Rather it acknowledges as an asset for all future times the main accidental outcome of the history of the market system, namely the creation of unprecedented processing capabilities which, if used correctly, can liberate mankind from the drudgery of laboring for the sake of sheer survival. Finally, in relation to the market system, my analysis even has a conservative overtone. It suggests that, purged of its unqualified expansionist trends and of its political and social abuses, the modern market may very well be the most viable and effective way devised to date to undertake mass production, the delivery of goods and services, and the organization of certain types of economizing social systems.

Any future which is envisioned as a linear development of the market-centered society will necessarily be worse than the present. Social science should be freed from its obsession with development and begin to understand that each contemporary society is potentially ready to become a good one if it chooses to rid itself of the linearist vision of history. This book suggests that there are many possibilities for the nations of the so-called underdeveloped world to recover immediately from their peripheral condition, if only they would find their own political will and thus free themselves from the syndrome of relative deprivation which they have internalized by taking the advanced market society as the scenario of their future.

Delimitation of the market system, as predicated by the *New Science,* implies the formulation and implementation of new allocative criteria and policies within and between nations. The novelty of these criteria mainly results from their sensitivity to the detrimental ecological and psychological externalities produced by the unregulated activities of the market system. The administration of constraints to the functioning of the market system aims at the preservation both of the ecological soundness of the planet and of the psychological health of mankind. Those constraints are to be discovered and invented through a complex research process subsumed under neither hegemonic interests nor doctrinary orthodoxies of any kind. For instance, socialism is extraneous to the para-economic model presented in this book. Indeed private initiative and private property are fundamental conditions for any successful delimitation of the market. But in a delimited society private initiative and private property are defended against the disguised power of privileged corporate actors as well as the omnipotent state. In fact the state has already been assigned this task which, in a delimited society, it would exercise more forcefully and systematically in the interest of a revitalizing diversification of social and communal life. More specifically, in the economic domain the delimitation of the market would entail, not the elimination of private investors, but the enforcement of governmental policies intended to guarantee the compatibility of the structure of production and the population's propensities to consume with ecological and socio-psychological requirements. This scenario does not imply socialism, i.e., ownership by the state of the instruments of production. It does, however, demand a redefinition of the goals and priorities according to which existing central state controls should be exercised.

My analysis suggests that, given the present possibility of producing an abundance of primal goods and services, the category of job now has limited utility for assessing an individual's social worth. Production is not necessarily an output of activities undertaken within the confines of the market. Rather it is constituted by outputs which contribute to enhancing the enjoyment of life, and therefore these outputs can be the result of activities undertaken within the confines of non-market-oriented social systems. In this sense resources are infinite and there are no limits to production. The obsession with job as the only criterion of human productive capacity is a fundamental blind spot of governmental policy makers and the conventional economics they employ. Institutional reforms as for instance implementation of an imaginative grant system, can be devised to reward multiple forms of an individual's productive contribution to social life, of which job constitutes only a particular form.

In the prevailing economic institutional framework, increasing job opportunities will require escalation of the production of demonstrative goods, but biophysical production constraints have precluded this strategy. As before the rise of the market-centered society, so today, in its historical decline, full employment of manpower is again possible without imposing upon all individuals willing to work the imperative of being a jobholder. Ignorance of this fact is particularly regrettable at a time when the economy is increasingly losing its capacity to provide jobs for all persons willing to work.

This book is nothing more than a preliminary theoretical statement of the new science of organizations. It simply sets a research agenda. Much is yet to be done to transform the *New Science* into a tool for social reconstruction. In its present terms my analysis did not, for instance, discuss how the state might systematically implement and manage delimited social systems. A state fit to monitor the kind of society envisioned by the *New Science*, although displaying regulatory characteristics, will not be a socialist intervenor. Rather it will be an authoritative convenor of social systems whose assignment is to guarantee their functional complementarity. How, institutionally, it will accomplish this role is a matter for further investigation. Moreover no operational guidelines for designing, implementing, maintaining, and linking the variety of complementary social systems were presented in this book. I assumed that to present such guidelines before articulating in theoretical terms the plight of the individual in the contemporary market-centered society would be pointless. I also assumed that the individual should first be freed from his psychological enthralment with the market mentality. Before these objectives are accomplished, any set of operational prescriptions would be of no avail to him. I am certainly aware of these and other omissions of this book, but I am already engaged in a further development of this analysis based upon what I am learning from the experiences of concerned people who are now, in many ways and places, striving to find viable alternatives to the present state of affairs of the world.

THE ENDURING ORGANIZATION

This book lays bare the fallacies of current organization theory, the demise of which is not to be regretted; it is, rather, an auspicious event.

Extant organization theory can no longer conceal its parochialism. It is parochial because it focuses upon organizational issues from the standpoint of criteria inherent in a type of society where the market plays the role of an overarching integrative pattern and force. It becomes speechless when challenged by organizational issues common to all societies. Moreover it is parochial because it feeds itself on the fallacy of simple location, i.e., ignorance of the interconnection and interdependence of things in the universe; it deals with things as if they were confined within mechanical sections of space and time.

In fairness there is much in extant organization theory which any alternative theory should appropriate and develop. More than ever we now have reasons to admit that a fundamental promise of the old theory can be delivered: the problem of poverty as a material condition can be solved technically. After all, the old theory taught us that knowledge can be utilized systematically to produce more, to produce better, to produce enough, while at the same time liberating men from laboring activities. It taught us that in the last analysis capital is processing capability; it is a verb, not a noun. But dupe of a narrow concept of production and capital, extant organization theory finds itself in a stalemate. We have learned that indefinitely increasing commodity production and unregulated technological advance are not necessarily conducive to actualization of man's potential. Within the confines of the dominant interests which have prevailed during the last three centuries, extant organization theory has already accomplished its assignment. Awareness of this fact paves the way for the elaboration of a multidimensional science of organizations.

The old theory implies that production is only a technical matter. However, the fundamental assumption of the new science of organizations is that production is both a technical and moral issue. Production is not only a mechanomorphic activity; it is also an outcome of men's creative enjoyment of themselves. In a sense men produce themselves while they produce things. In other words production should be undertaken not only to provide enough goods for man to live a healthy life, but to provide the conditions for him to actualize his nature and enjoy doing so. Thus the production of commodities must be managed ethically, because as an infinite consumer man does not endure but exhausts his very being. Moreover production is also a moral issue because of its impact upon nature at large. Indeed nature is not inert material; it is a living system which can only endure as long as one does not violate the biophysical constraints superimposed upon its restorative processes.

The use of the verb 'to endure' in the preceding paragraph is intentional. Endurance is a category of physical, human, and social existence. Apart from endurance one cannot understand the process through which things, human beings, and societies accomplish their inherent individualities. Thus endurance does not imply maintenance. It is retention of character through change; it is victory over fluidity. It is a category of process thinking which acknowledges that everything is interconnected and continuously striving towards an optimal balance between conservation and change in the process of attaining a patterned achievement of its intrinsic ends.[1] Characterizing the general meaning of endurance, Whitehead writes: 'Endurance is the retention through time of an achievement of value. What endures is identity of pattern, self-inherited. Endurance requires the favorable environment. The whole of science revolves round the question of enduring organisms' (Whitehead 1967: 194). This citation sets the scenario for the elucidation of the parochialisms characteristic of extant organization theory.

Conventional organization theory and social science in general are not prone to acknowledge the viability of non-Western societies on their own value grounds. In the conceptual perspective of these disciplines, the Westernization of those societies is tantamount to their qualitative enhancement. For instance, this ideological bias is clearly spelled out by Likert (1963). The concepts and principles of what he calls a 'world-wide theory of management' are all deduced from the practice of Western industrial experience. Likert explicitly justifies the 'universality' of these concepts and principles, though not properly in theoretical terms, since in his view his doctrine is not fit to manage resources in any context, but only in Westernized industrial sectors. He considers such a doctrine universal because it is predicated upon Western expansionism which is narrowing 'the cultural differences ... among nations' and making them 'much more alike in their organization (and industrial) existence.' Such theorizing is completely insensitive to dramatic facts which demonstrate that the Western model of industrialization is upsetting the organizational basis of peripheral societies rather than enhancing their capabilities to endure as self-determinative systems. A truly universal organization theory cannot indulge in such historical parochialism. Rather it should imply that the search for organizational requirements constitutes a concrete matter in each society and defies 'concepts' and 'principles' as Likert conceives them. Christopher Alexander correctly envisions such a search as an analytical process leading to the discovery and implementation of a good 'fit' enabling the satisfaction of the mutual demands which context and form make on one another (Alexander 1974: 19). Alexander is suggesting an endurance-oriented process of design. He therefore acknowledges that 'in their own way the simple cultures do their job better than we do ours' (32).

The disruption of enduring life systems is also a current feature of Western industrialized societies.[2] The practice of organization design which prevails in these countries is largely affected by the fallacy of simple location. Much of Georgescu-Roegen's thermodynamic analysis of conventional economic thinking reveals the fallacy of simple location. Organizations and the economic process that they enforce are usually conceived as if they had no connections with the biophysical sphere. Such a conception neglects the fact that the economic process, and especially the type of organization designed according to purely economizing criteria, continuously draws from the environment low entropy matterenergy and returns it in a state of high entropy. In this process the environment is necessarily depleted and polluted and thus the conditions required for enduring physical, human, and social existence are upset. Postulated upon the fallacy of simple location, extant organization theory is, rather, prone to aggravate the increasing thermodynamic unbalance plaguing Western societies. The time has come to replace it with an endurance-centered science of organizations.

It should now be clear to the reader that in one sense the *New Science of Organizations* is not really new[3] for it is as old as common sense. What is new are the circumstances under which we must once again begin to listen to our innermost selves.

Notes and References

CHAPTER 1

Notes

1 See Parsons and Shils 1962; Parsons 1964.
2 On the history of the Frankfurt school, see Jay 1973.
3 Horkheimer writes: 'Man has gradually become less dependent upon absolute standards of conduct, universally binding ideas. He is held to be so completely free that he needs no standards except his own. Paradoxically, however, this increase of independence has led to a parallel increase of passivity. Shrewd as man's calculations have become as regards his means, his choice of ends, which has formerly correlated with belief in an objective truth, has become witless: the individual, purified of all remnants of mythologies, including the mythology of objective reason, reacts automatically, according to general patterns of adaptation. Economic and social forces take on the character of blind natural powers that man, in order to preserve himself, must dominate by adjusting himself to them. As the end result of the process, we have on the one hand the self, the abstract ego emptied of all substance except its attempt to transform everything in heaven and on earth into means of its preservation, and on the other hand, an empty nature degraded to mere material, mere stuff to be dominated, without other purpose than that of his very domination' (Horkheimer 1947:97).
4 See Jay 1973: 262.
5 See especially Horkheimer 1947: 141–2.
6 Cited in Habermas 1968: 200.
7 See Burke 1963/4.

8 See Habermas 1970: 116–46.
9 Cited in Sandoz 1971: 59.
10 Cited in Voegelin 1967: 151.
11 See Voegelin 1974.

References

Burke, K. 1963/4. 'Definition of Man,' *Hudson Review*, winter
Habermas, Jürgen. 1970. *Toward a Rational Society: Student Protest, Science and Politics*. Boston: Beacon Press
– 1970a. 'Toward a Theory of Communicative Competence,' in H. P. Drietzel ed, *Recent Sociology, no. 2*. New York: Macmillan
– 1971. *Knowledge and Human Interests*. Boston: Beacon Press
– 1973. *Theory and Practice*. Boston: Beacon Press
Hobbes, Thomas. 1839. *The English Works*, vol. I. Molesworth Edition. London: John Bohn
– 1974. *Leviathan*. Edited by Michael Oakeshott with an introduction by R.S. Peters. London: Macmillan
Horkheimer, Max. 1947. *Eclipse of Reason*. New York: Oxford University Press
Horkheimer, Max and Adorno, Theodor W. 1972. *Dialectic of Enlightenment*. New York: Herder and Herder
Husserl, Edmund 1965. *Phenomenology and the Crisis of Philosophy*. New York: Harper and Row
– 1967. 'The Thesis of the Natural Standpoint and Its Suspension,' in J. Kockelmans ed, *Phenomenology*. Garden City, NY: Doubleday
Jay, Martin. 1973. *The Dialectical Imagination: A History of the Frankfurt School and the Institute of Social Research*. Boston: Little, Brown
Keynes, J.M. 1932. 'Economic Possibilities for Our Children,' in Keynes, *Essays in Persuasion*. New York: Harcourt, Brace
Mannheim, Karl. 1939. *Ideology and Utopy*. New York: Harcourt, Brace and World
– 1940. *Men and Society in an Age of Reconstruction*. New York: Harcourt, Brace and World
Parsons, Talcott. 1964. 'Evolutionary Universals in Society,' *American Sociological Review*, June
Parsons, Talcott and Shils, E.A. eds. 1962. *Toward a General Theory of Action*. New York: Harper and Row
Sandoz, Ellis. 1971. 'The Foundations of Voegelin's Political Theory,' *The Political Science Reviewer* 1, fall
Voegelin, Eric. 1952. *The New Science of Politics*. Chicago: University of Chicago Press

- 1960. 'El Concepto de la "Buena Sociedad,"' *Cadernos del Congresso por la Libertad*. Suplemento del no. 40
- 1961. 'On Readiness to Rational Debate,' in A. Hynold ed, *Freedom and Serfdom*. Dordrecht: D. Reidel Publishing
- 1963. 'Industrial Society in Search of Reason,' in R. Aron ed, *World Technology and Human Destiny*. Ann Arbor: University of Michigan Press
- 1967. 'On Debate and Existence,' *The Intercollegiate Reviewer*, March/April
- 1974. 'Reason: The Classic Experience,' *The Southern Review*, spring
Weber, Max. 1968. *Economy and Society*, vol. 1. New York: Bedminster Press
- 1969. *Methodology of Social Sciences*. New York: The Free Press

CHAPTER 2

Notes

1 Eric Voegelin and Leo Strauss have also highlighted the contradictions of Weber's idea of value-neutral social science in Voegelin 1952: 13–22 and Strauss 1953: 35–80.
2 On the essentialist and nominalist character of concepts, see Popper 1971: 32, 1965: 13–14.
3 On the complex controversy surrounding substantive economic theory versus formal economic theory see Kaplan 1968 and Cook 1966.
4 The 'embeddedness' of the economy in society is one of Karl Polanyi's overarching ideas. See Dalton 1971.
5 On the historical process of 'sublimation' of the political element, see Wolin 1960.
6 On the transvaluation of sociality and the classical political thought, see Arendt 1958.
7 On the influence of Mandeville on Adam Smith, see Marx 1974: 355, Colletti 1972, and Macfie 1967.
8 See Bryson 1945 and Stephen 1927.
9 See Bryson 1945: 118.
10 On this point, see Myrdal 1954. He states that 'the idea of economy as a kind of social housekeeping inspires not only the theory of free trade but all other doctrines of economic policy' (140).
11 See Barker 1959: 357.
12 On the concept of the political, see Aristotle's *Politics*.
13 See Dijksterhuis 1969: 72–5.
14 Quoted in Flynn 1942: 23. Flynn's translation of Saint Thomas was lightly modified by the author.

15 See Sewall 1901: 38–43.
16 'When men are thus dependent on one another and reciprocally related to one another in their work and the satisfaction of their needs, subjective self-seeking turns into a contribution to the satisfaction of the needs of everyone else. That is to say, by a dialectical advance, subjective self-seeking turns into the mediation of the particular through the universal, with the result that each man in earning, producing and enjoying on his own account is *eo ipso* producing and earning for the enjoyment of everyone else' (Hegel 1973: 129–30).
17 On historical breakthroughs see Voegelin 1968.
18 See Heisenberg 1975: 56.
19 Cited in Eugene F. Miller, 'David Easton's Political Theory,' *The Political Science Reviewer* (Fall 1971).
20 Under such prevailing orientation it is not surprising to find the following statements in an influential treatise in political theory: ' "What is good for General Motors is good for the country" contains at least a partial truth. "What is good for the Presidency is good for the country" contains, however, more truth ... The power of the presidency is identified with the good of the polity' (Huntington 1968: 26). Apparently Almond is now realizing the conceptual flaws of conventional political science: see Almond and Genco 1977.
21 This statement is derived from Sartori (1970). He writes: 'A few years ago Karl Deutsch predicted that by 1975 the informational requirements of political science would be satisfied by some "fifty million card-equivalents [of IBM cards] ... and a total annual growth of perhaps as much as five million." I find the estimate frightening.' (Sartori 1970: 1035–6).

References

Almond, Gabriel A. 1973. 'Slaying the Functional Dragon: A Reply to Stanley Rothman,' *The Political Science Reviewer*, fall
Almond, G.A. and Genco, S.J. 1977. 'Clouds, Clocks and the Study of Politics,' *World Politics* 29, no. 4, July
Arendt, H. 1958, *The Human Condition*. Garden City, NY: Doubleday
Aristotle. 1972. *Politics*. Translated with an introduction, notes and appendixes by Ernest Barker. Oxford: Oxford University Press
Barker, Ernest. 1959. *The Political Thought of Plato and Aristotle*. New York: Dover Publications
Bryson, Gladys. 1945. *Man and Society. The Scottish Inquiry of the Eighteenth Century*. Princeton, NJ: Princeton University Press
Clark, J.M. 1960. *Alternative to Serfdom*. New York: Vintage Books
Colletti, Lucio. 1972. *From Rousseau to Lenin: Studies in Ideology and Society*. New York: Monthly Review Press

Cook, Scott. 1966. 'The Obsolete Anti-Market Mentality: A Critique of the Substantive Approach to Economic Anthropology,' *American Anthropologist* 58

Cropsey, Joseph. 1957. *Polity and Economy: An Interpretation of the Principles of Adam Smith*. The Hague: Martinus Nijhoff

Dalton, George ed. 1971. *Primitive, Archaic and Modern Economies*. Essays of Karl Polanyi. Boston: Beacon Press

Daly, Herman E. ed. 1973. *Toward a Steady-State Economy*. San Francisco: W.H. Freeman

Deutsch, K.W. 1960. *The Nerves of Government*. New York: The Free Press

Dijksterhuis, E.J. 1909. *The Mechanization of the World Picture*. Oxford: Oxford University Press

Easton, David. 1973. 'Systems Analysis and the Classical Critics,' *The Political Science Reviewer*, fall

Eddington, Arthur. 1974. *The Nature of the Physical World*. Ann Arbor: University of Michigan Press

Flynn, Frederick E. 1942. *Wealth and Money in the Economic Philosophy of St. Thomas*. Notre Dame, Ind.: University of Notre Dame Press

Galbraith, J.K. 1970. 'Economics as a System of Belief,' *American Economic Review*, May

Georgescu-Roegen, Nicholas. 1975. 'The Entropy Law and the Economic Problem,' in Daly 1973

Hegel, G.W.F. 1973. *Philosophy of Right*. Translated with notes by T.M. Knox. New York: Oxford University Press

Heisenberg, Werner. 1975. 'The Great Tradition, End of an Epoch?' *Encounter*, March

Hume, David. 1973. *A Treatise of Human Nature*. Edited by L.A. Selby-Bigge. Oxford: Oxford University Press

Huntington, Samuel P. 1968. *Political Order in Changing Societies*. New Haven, Conn.: Yale University Press

Jonas, Hans. 1969. 'Economic Knowledge and Critique of Goals,' in R.L. Heilbroner ed, *Economic Means and Social Ends: Essays in Political Economy*. Inglewood Cliffs, NJ: Prentice-Hall

Kaplan, David. 1968. 'The Formal-Substantive Controversy in Economic Anthropology: Reflections on Its Implications,' *Southwestern Journal of Anthropology*, autumn

Keynes, J.M. 1932. *Essays in Persuasion*. New York: Harcourt, Brace

Lonergan, Bernard. 1967. *Collection*. New York: Herder and Herder

Lowe, Adolphe. 1967. 'The Normative Roots of Economic Value,' in Sidney Hook, *Human Values and Economic Policy*. New York: New York University Press

MacFie, A.L. 1967. *The Individual in Society: Papers on Adam Smith*. London: George Allen and Unwin

Macpherson, C.B. 1962. *The Political Theory of Possessive Individualism*. Oxford: Oxford University Press
– 1973. *Democratic Theory*. Oxford: Oxford University Press
Martin, David A. 1973. 'Beyond Positive Economics: Towards Moral Philosophy,' *The American Economist*, spring
Marx, Karl. 1974. *Capital*, vol. I. New York: International Publishers
Mill, John Stuart. 1857. *Principles of Political Economy*, vol. II. London: John W. Parker and Son
Morrow, Glenn R. 1923. *The Ethical and Economic Theories of Adam Smith*. New York: Longmans, Green
Popper, Karl R. 1971. *The Open Society and Its Enemies*, vol. I. Princeton, NJ: Princeton University Press
– 1963. *The Open Society and Its Enemies*, vol. II. New York: Harper and Row
Sartori, Giovanni. 1970. 'Concept Misformation in Comparative Politics,' *The American Political Science Review*, December
Schneider, Herbert W. 1970. *Adam Smith's Moral and Political Philosophy*. New York: Harper and Row
Sewall, H.R. 1901. *The Theory of Value before Adam Smith*. Publications of the American Economic Association, Third Series, vol. II, no. 3
Smith, Adam. 1970. *Theory of Moral Sentiments*. Edited by H.W. Schneider and constituting part of Schneider 1970
– n.d. *An Inquiry into the Nature and Causes of the Wealth of Nations*. London: Ward, Lock and Co., Warwick House
Stephen, Leslie. 1927. *History of English Thought in the Eighteenth Century*. 2 vols. London: John Murray
Strauss, Leo. 1953. *Natural Right and History*. Chicago: University of Chicago Press
Voegelin, Eric. 1952. *The New Science of Politics*. Chicago: University of Chicago Press
– 1968. 'Configurations of History,' in P.G. Kuntz ed, *The Concept of Order*, Seattle, Wash.: University of Washington Press
Von Stein, Lorenz, 1964. *The History of the Social Movement in France, 1789–1850*. Totowa, NJ: Bedminster Press
Von Treitschke, Heinrich 1963. *Politics*. Abridged, edited, and with an introduction by Hans Kohn. New York: Harcourt, Brace and World
Weber, Max. 1958. *The Protestant Ethic and the Spirit of Capitalism*. New York: Charles Scribner's Sons
Wicksteed, Philip W. 1913. 'The Scope and Method of Political Economy in the Light of the "Marginal" Theory of Distribution,' in R.L. Smyth ed, *Essays in Economic Method*. London: Gerald Duckworth, 1952
Wolin, Sheldon S. 1960. *Politics and Vision*. Boston: Little, Brown

CHAPTER 3

Notes

1 Thus, as Reichenbach puts it, ethics would only 'inform us about matters of fact. Of this kind is a descriptive ethics which informs us about the ethical habits of various people and social classes; such an ethics is a part of sociology, but is not of a normative nature' (Reichenbach 1959: 276–7).
2 On representation and its metahistorical grounds, see Voegelin 1952 and Eliade 1959.
3 Hirschman (1977) addresses the question of *interests* as related to the rise of capitalism.
4 Correctly, L. von Stein points out that modern society is constituted when the 'organization of economic life becomes the order of human community' (von Stein 1964: 47). Therefore, 'the awareness of [interest] ... regulates all outward activities ... is always present and alive in every individual, determining his social position' (55).
5 On this point, see Hauser 1965 and Lowenthal 1968.
6 On the artistic market in Italy, and the rise of modern art, see von Martin 1944.
7 On the 'Machiavellian moment,' see Pocock 1975.
8 Cited in Sjoberg 1959: 605.
9 Cited in Sjoberg 1959: 606.
10 A further development of this point exceeds the scope of this chapter. See, however, Capek 1961 and Leclerc 1972.

References

Aristotle. 1972. *Politics*. Edited and translated by Ernest Barker. Oxford: Oxford University Press
Bacon, Francis. 1968. *The New Organon*. Vol. IV of Francis Bacon, *Works*. Edited by J. Spedding et al. New York: Garret Press
Capec, Milic. 1961. *Philosophical Impact of Contemporary Physics*. Princeton, NJ: D. Van. Nostrand
Castiglione, Baldesar. 1976. *The Book of the Courtier*. Middlesex, England: Penguin Books
Cicero. 1975. *De Officis*. Cambridge, Mass.: Harvard University Press
Eliade, Mircea. 1959. *Cosmos and History*. New York: Harper & Row
Hauser, Arnold. 1965. *Mannerism*. London: Routledge & Kegan Paul
Hirschman, Albert O. 1977. *The Passions and the Interests*. Princeton: NJ: Princeton University Press

Hobbes, Thomas. 1839. *The English Works*, vol. I. Molesworth Edition
- 1840. *The English Works*, vol. IV. Molesworth Edition
- 1841. *The English Works*, vol. V. Molesworth Edition
Leclerc, Ivor. 1972. *The Nature of Physical Existence*. London: George Allen & Unwin
Lowenthal, Leo 1968. *Literature, Popular Culture, and Society*. Palo Alto, Calif.: Pacific Books
Machiavelli, N. 1965. *The Prince*. In *Machiavelli, The Chief Works and Others*, vol. I. Translated by Allan Gilbert. Durham, NC: Duke University Press
Montaigne, Michel Eyquem de 1975. *The Complete Essays of Montaigne*. Translated by D.M. Frame. Stanford, Calif.: Stanford University Press
Pocock, J.G.A. 1975. *The Machiavellian Moment*. Princeton, NJ: Princeton University Press
Reichenbach, Hans 1959. *The Rise of Scientific Philosophy*. Berkeley: University of California Press
Sjoberg, G. 1959. 'Operationalism and Social Research,' in L. Gross ed, *Symposium on Sociological Theory*. New York: Harper & Row
Smith, Adam, 1976. *The Theory of Moral Sentiments*. Indianapolis, Ind.: Liberty Classics
Voegelin, Eric. 1952. *The New Science of Politics*. Chicago: University of Chicago Press
Von Martin, Alfred. 1944. *Sociology of the Renaissance*. New York: Oxford University Press
Von Stein, Lorenz. 1964. *The History of the Social Movement in France, 1789–1850*. Totowa, NJ: The Bedminster Press
Whitehead, A.N. 1967. *Science and the Modern World*. New York: The Free Press
- 1969. *Process and Reality*. New York: The Free Press
Wolin, Sheldon S. 1960. *Politics and Vision*. Boston: Little, Brown

CHAPTER 4

Notes

1 See Donald A. Schon 1963. On displacement of concepts as a tool of technological innovation, see W.J.J. Gordon 1973.
2 See Ernest Nagel 1961: 108. Nagel writes: 'Similarities between the new and old are often only vaguely apprehended without being carefully articulated. Moreover, little if any attention is generally paid to the limits within which

such felt resemblances are valid. Accordingly, when familiar notions are extended to novel subject matters on the basis of unanalyzed similarities, serious errors can easily be committed' (108).

3 About the fallacies of what Giovanni Sartori calls 'conceptual stretching' see Sartori 1970.

4 See, for instance, Beatrice and Sydney Rome 1967: 181.

5 Cited in Richard Means 1970: 173.

6 The concept estrangement (or alienation), as Paul Tillich states, corresponds to 'what in religious symbolism is called the fall.' See Tillich 1968: 419.

7 Marx, for instance, states: 'Theology explains the origin of evil by the fall of man; that is, it asserts as a historical fact what it should explain' (Marx 1964: 121).

8 See Shepard 1972, Kirsch and Lengermann 1972, and Miller 1975.

9 Echoing Bennis, H.M.F. Rush writes: 'The organization is believed to have ... all the qualities of an individual' (Rush 1969: 8).

10 On the notions of partial inclusion and total inclusion, see F.H. Allport 1933.

11 See Chapter 3 of this book.

12 See Glass 1972, 1974 and Laing 1967, 1969.

13 See for instance, Bertram Gross 1973; David K. Hart and William G. Scott 1975; James Glass 1975; C. Perrow 1972; E. Singer and M. Wooton 1976; and W.N. Dunn and B. Fozouni 1976.

14 See Eric Voegelin 1956.

15 See an excellent discussion about this issue in J.L. Esposito 1976. See also J. Weizenbaum 1976.

16 On the 'fallacy of misplaced participation' characteristic of the human relations school, see Riesman 1971: 270–1. See Whyte 1957: 38–42 for the critique of the Hawthorne experiment.

17 See Ramos 1972.

18 See Caplow 1964: 234. The next item is largely drawn from Caplow's book, same page.

References

Allport, F.H. 1933. *Institutional Behavior*. Chapel Hill: University of North Carolina Press

Bennis, W. 1966. *Changing Organizations*. New York: McGraw-Hill

Blauner, R. 1964. *Alienation & Freedom*. Chicago: University of Chicago Press

Boguslaw, R. 1965. *The New Utopians*. Englewood Cliffs, NJ: Prentice-Hall

Caplow, T. 1964. *Principles of Organization*. New York: Harcourt, Brace & World

Cassirer, E. 1951. *The Philosophy of Enlightenment*. Princeton: Princeton University Press

Deutsch, K. 1966. *The Nerves of Government*. New York: The Free Press

Drucker, P. 1969. *The Age of Discontinuity*. New York: Harper & Row

Dunn, W.N. and Fozouni, B. 1976. *Toward a Critical Administrative Theory*. Beverly Hills, Calif.: Sage Publications

Esposito, J.L. 1976. 'System, Holons, and Persons: A Critique of Systems Philosophy,' *International Philosophical Quarterly*, June

Glass, J. 1972. 'Schizophrenia and Perception, A Critique of the Liberal Theory of Externality,' *Inquiry* 15

– 1974. 'Epicurus and the Modern Culture of Withdrawal,' *American Politics Quarterly* 2, no. 2, July

– 1975. 'Consciousness and Organization,' *Administration and Society*, November

Gordon, W.J.J. 1973. *Synectics*. New York: Macmillan

Gross, B. 1973. 'An Organized Society,' *Public Administration Review*, July/August

Gurvitch, G. 1958. 'Microsociologie,' in G. Gurvitch ed, *Traité de Sociologie*. Paris: Presses Universitaires de France

Habermas, Jürgen. 1970. *Toward a Rational Society*. Boston: Beacon Press

Hart, D.K. and Scott, W.G. 1975. 'The Organizational Imperative,' *Administration and Society*, November

Kant, Immanuel. 1965. *The Metaphysical Elements of Justice*. New York: Bobbs-Merril

Kaplan, A. 1964. *The Conduct of Inquiry*. San Francisco: Chandler Publishing.

Katz, D. and Kahn, R.L. 1966. *The Social Psychology of Organizations*. New York: John Wiley & Sons

Kirsch, B.A. and Lengermann, J.J. 1972. 'An Empirical Test of Blauner's Ideas on Alienation in Work as Applied to Different Jobs in a White Collar Setting,' *Sociology and Social Research* 56, January

Koestler, A. 1969. 'Beyond Atomism and Holism – The Concept of Holon,' in A. Koestler and J.J. Smythies eds, *Beyond Reductionism*. New York: Macmillan

– 1977. 'Free Will in a Hierarchic Context.' in J.B. Cobb and D.R. Griffin eds, *Mind in Nature*. Washington: University Press of America

Laing, R.D. 1967. *The Politics of Experience*. New York: Ballantine Books

– 1969. *The Divided Self*. London: Penguin Books

Luthans, F. 1977. *Organizational Behavior*. New York: McGraw-Hill

Marx, Karl. 1964. *Karl Marx: Early Writings*. Edited by T.B. Bottomore. New York: McGraw-Hill

Means, R. 1970. *The Ethical Imperative*. Garden City, NY: Doubleday

Merton, R.K. 1967. *Social Theory and Social Structure*. New York: The Free Press

Miller, J. 1975. 'Isolation in Organizations: Alienation from Authority, Control and Expressive Relations,' *Administrative Science Quarterly*, June

Mooney, J.D. and Reiley, A.C. 1939. *The Principles of Organization*. New York: Harper & Brothers

Nagel, E. 1961. *The Structure of Science: Problems in the Logic of Scientific Explanation*. New York: Harcourt, Brace & World

Perrow, C. 1972. *Complex Organization: A Critical Essay*. Glenview, Illinois: Scott, Foresman

Ramos, A.G. 1972. 'Models of Man and Administrative Theory,' *Public Administration Review*, May/June

Riesman, D. et al. 1971. *The Lonely Crowd*. New Haven, Conn.: Yale University Press

Rome, Beatrice and Sydney. 1967. 'Humanistic Research on Large Organizations,' in J.F.T. Bugental ed, *Challenges of Humanistic Psychology*. New York: McGraw-Hill

Rush, H.M.F. 1969. *Behavioral Science, Concepts and Management Applications*. New York: National Industrial Conference Board

Sartori, G. 1970. 'Concept Misformation in Comparative Politics,' *The American Political Science Review* 64, December, no. 4

Schacht, R. 1970. *Alienation*. Garden City, NY: Doubleday

Schon, D. 1963. *Displacement of Concepts*. London: Tavistock Publications

Seeman, M. 1972. 'On the Meaning of Alienation,' in Ada W. Finifter ed, *Alienation and the Social System*. New York: John Wiley & Sons

Shepard, J.M. 1972. 'Alienation as a Process: Work as a Case in Point,' *The Sociological Quarterly* 13, no. 2

Singer, E. and Wooton, M. 1976. 'The Triumph and Failure of Albert Speer's Administrative Genius: Implications for Current Management Theory and Practice,' *The Journal of Applied Behavioral Science* 12, no. 1

Tillich, P. 1968. *History of Christian Thought*. Edited by C.E. Bracten. New York: Simon and Schuster

Voegelin, Eric. 1956. 'Necessary Moral Bases for Communication in a Democracy,' in R.C. Seitz ed, *Problems of Communication in Pluralistic Society*. Milwaukee, Wis.: Marquette University Press

Walton, R.E. 1972. 'How to Counter Alienation in the Plant,' *Harvard Business Review*, November-December

Weizenbaum, J. 1976. *Computer Power and Human Reason*. San Francisco: W.N. Freeman

Whyte, W.H. 1975. *The Organization Man*. Garden City, NY: Doubleday

Wolin, S.S. 1969. 'Political Theory as a Vocation,' *The American Political Science Review* 63, December, no. 4

CHAPTER 5

Notes

1 See Voegelin 1963: 36. The notion of 'civil theology' has a long tradition in the field of political theory. For an excellent overview of this notion, see Ellis Sandoz 1972. See also Dante Germino 1967: 29. An attempt at formulating a civil theology for advanced industrial society is represented by J. Rawls 1971. In spite of the compassionate flavor of Rawls's notion of social equity, which is becoming influential among some administrative theorists, his 'theory of justice' in the last analysis reflects a myopic assessment of the present state of advanced industrial society, to the extent that the rationality inherent in the market pricing system is explicitly accepted by Rawls as a basic assumption of his 'moral philosophy.'

2 See also McClelland and Winter 1969.

3 In the England of the Tudors and the Stuarts, says Peter Laslett, 'adults did not go out to work' and 'institutional life was almost unknown' (Laslett 1965: 11).

4 Supportive data for this statement is found, for instance, in the works of Thurnwald, Malinowski, Firth, Lowie, and Radcliffe-Brown. See Polanyi 1971 and Sahlins 1972.

5 'We economists,' said Milton Friedman in 1972, 'have claimed more than we can deliver.' Actually he echoed Arthur F. Burns, Former chairman of the Federal Reserve Board, who in 1971 said, 'The rules of economics are not working quite the way they used to.' Cited in Henderson 1978: 63–4.

6 See, for example, chapter 3, 'Toward the Organizational Society' in Presthus 1965. See also Scott and Hart 1979. Epistemological dimensions of the organizational society are discussed by Haworth 1977.

7 Motivation research projects are widely encouraged in this period. Whyte quotes Ernest Dichter, a 'motivation researcher': 'We are now confronted with the problem of permitting the average American to feel moral even when he is flirting, even when he is spending, even when he is not saving, even when he is taking two vacations a year and buying a second or third car. One of the basic problems of this prosperity, then, is to give people the sanction and justification to enjoy it and to demonstrate that the hedonistic approach to his life is a moral, not an immoral one' (Whyte 1957: 19). Leo Lowenthal's 'Triumph of Mass Idols' provides an interesting study on how the transition from a production-oriented economy to a consumption-oriented economy was reflected in American popular culture 1968.

8 See Galbraith 1970.

9 See, for instance, Argyris 1973, 1973a, and 1973b.

10 See de Grazia 1964: 51.
11 See Schumpeter 1974: 270. See also Garraty 1978.
12 One should realize that this distinction intends to avoid the utilization of the classification of productive activities as primary, secondary, and tertiary which is current in economic textbooks. My notion of demonstrative goods and services reflects Duesemberry's concept of 'demonstration effect,' although it has similarities with what Fred Hirsch (1976) calls 'positional' goods.
13 This and other quotations of Mill are drawn from Daly 1973: 12–13.
14 Quoted in Heilbroner 1972: 241.
15 See the exchange between Simon (1973) and Argyris (1973).
16 Solid psychological scholarship would warrant this statement. One should not forget that Freud himself stated in *Civilization and Its Discontents* that 'in the severity of its commands and prohibitions [the super-ego] troubles itself too little about the happiness of the ego' (Freud 1962: 90). But to say the least Freud was ambiguous about the irreductibility of the self to sociality. Better sources to substantiate the statement would be found in writings of Carl Jung, Alfred Adler, Otto Rank, Franz Alexander, H. Hartmann, Wilhelm Stekel, L. Binswanger, Erich Fromm, M. Boss, Viktor Frankl, R.D. Laing, Ira Progoff, Rollo May, and others. The substantiation of the statement cannot be made in the text of this chapter, lest it go astray of its main objective.
17 See Raskin 1973: xxiii.

References

Argyris, Chris. 1973. 'Organization Man: Rational *and* Self-actualizing,' *Public Administration Review*, July/August
– 1973a. 'Some Limits of Rational Man Organizational Theory,' *Public Administration Review*, May/June
– 1973b. *On Organizations of the Future*. Beverly Hills, Calif. Sage Publications
Aristotle. 1954. *Rhetoric and Poetics*. Translated, respectively, by W.R. Roberts and I. Baywater. New York: Modern Library
Barnard, C.I. 1948. *The Functions of the Executive*. Cambridge, Mass.: Harvard University Press
Daly, H.E. ed. 1973. *Toward a Steady-State Economy*. San Francisco: W.H. Freeman
de Grazia, Sebastian. 1964. *Of Time, Work, and Lesiure*. Garden City, NY: Doubleday
Deutsch, Karl. 1953. *Nationalism and Social Communication*. Cambridge, Mass.: MIT Press
Freud, Sigmund. 1962. *Civilization and Its Discontents*. New York: W. E. Norton

Galbraith, J.K. 1970. 'Economics as a System of Belief,' *American Economic Review,* May

Garraty, J.A. 1978. *Unemployment in History: Economic Thought and Public Policy.* New York: Harper & Row

Germino, Dante, 1967. *Beyond Ideology: The Revival of Political Theory.* New York: Harper and Row

Harrington, A. 1959. *Life in the Crystal Palace.* New York: Alfred A. Knopf

Haworth, L. 1977. *Decadence and Objectivity.* Toronto: University of Toronto Press

Heilbroner, R.L. 1972. *The Worldly Philosophers.* New York: Simon and Schuster

Henderson, Hazel. 1978. *Creating Alternative Futures.* New York: Berkley Publishing

Hirsch, Fred. 1976. *Social Limits to Growth.* Cambridge, Mass.: Harvard University Press

Inkeles, A. 1960. 'Industrial Man: The Relation of Status to Experience, Perception and Values,' *The American Journal of Sociology,* July

Jourard, S.M. 1964. *The Transparent Self.* Princeton, NJ: D. Van Nostrand

Keynes, J.M. 1964. *The General Theory of Employment, Interest, and Money.* New York: Harcourt, Brace & World

Laing, R.D. 1968. *The Politics of Experience.* New York: Ballantine Books

Laslett, Peter. 1965. *The World We Have Lost.* New York: Charles Scribner's Sons

Lazarsfeld, P.F. and Merton, R.K. 1974. 'Mass Communication, Popular Taste, and Organized Social Action,' in W. Schramm and D.F. Roberts eds, *The Process and Effects of Mass Communications.* Urbana: University of Illinois Press

Leiss, W. 1976. *The Limits to Satisfaction.* Toronto and Buffalo: University of Toronto Press

Lerner, D. 1958. *The Passing of Traditional Society: Modernizing the Middle East.* Glencoe, Ill.: The Free Press

Lowenthal, Leo. 1968. *Literature, Popular Culture and Society.* Palo Alto, Calif.: Pacific Books

Mannheim, K. 1940. *Man and Society in an Age of Reconstruction.* New York: Harcourt, Brace and World

McClelland, D.C. 1961. *The Achieving Society.* New York: D. Van Nostrand

McClelland, D.C. and Winter, D.G. et al. 1969. *Motivating Economic Achievement.* New York: The Free Press

Merton, R.K. 1967. *Social Theory and Social Structure.* New York: The Free Press

O'Toole, J. et al. 1973. *Work in America.* Washington, DC: The Colonial Press

Parsons, Talcott. 1964. 'Evolutionary Universals in Society,' *American Sociological Review,* June

Perk, H.F.W. 1966. 'The Great Transformation,' *The American Scholar* 35, spring, no. 2

Pfiffner, J. and Sherwood, F. 1965. *Administrative Organization.* Englewood Cliffs, NJ: Prentice-Hall

Plato. 1971. *Gorgias*. Translated by W.D. Woodhead. In Plato, *The Collected Dialogues*. Edited by E. Hamilton and H. Cairns. Bollinger Series LXXI. Princeton, NJ: Princeton University Press

Polanyi, Karl. 1971. 'Societies and Economic Systems,' in G. Dalton ed, *Primitive, Archaic and Modern Economies: Essays of Karl Polanyi*. Boston: Beacon Press

Polybius. 1972. *The Histories*, vol. III. Cambridge, Mass.: Harvard University Press

Presthus, R. 1965. *The Organizational Society*. New York: Random House

Raskin, M. 1973. *Being and Doing*. Boston: Beacon Press

Rawls, John. 1971. A Theory of Justice. Cambridge, Mass.: Harvard University Press

Sahlins, Marshall. 1972. *Stone Age Economics*. Chicago: Aldine-Atherton

Sandoz, Ellis. 1972. 'The Civil Theology of Liberal Democracy: Locke and His Predecessors,' *The Journal of Politics*, February

Schumpeter, J.A. (1974). *History of Economic Analysis*. New York: Oxford University Press

Scott, W.G. and Hart, D.K. 1979. *Organizational America*. Boston: Houghton, Mifflin

Simon, H.A. 1965. *Administrative Behavior*. New York: The Free Press

– 1969. *The Sciences of the Artificial*. Cambridge, Mass.: MIT Press

– 1973. 'Organization Man: Rational or Self-actualizing?' *Public Administration Review*, July/August

Smith, Adam. 1965. *An Inquiry into the Nature and Causes of the Wealth of Nations*. New York: Modern Library

Townsend, R. 1970. *Up the Organization: How to Stop the Corporation from Stifling People and Strangling Profits*. Greenwich, Conn.: Fawcett Publications

Voegelin, Eric. 1963. 'Industrial Society in Search of Reason,' in R. Aron ed, *World Technology and Human Destiny*. Ann Arbor: University of Michigan Press

Weizenbaum, J. 1976. *Computer Power and Human Reason*. San Francisco: W. H. Freeman

Whyte, Jr., W.H. 1957. *The Organization Man*. Garden City, NY: Doubleday

Wolin, S. 1969. 'Political Theory as a Vocation,' *The American Political Science Review*, December

CHAPTER 6

Notes

1 Simon states: 'it is impossible for the behavior of a single isolated individual to reach any degree of rationality' (1965: 79). He clearly indicates that only corporate actors and organizations behave rationally. Thus, Simon writes:

'[the] organization permits the individual to approach reasonably near to objective rationality' (80).

2 It is just such an erroneous assumption which pervades the polemic between Simon and Argyris. See Argyris 1973a and 1973b. See also Simon 1973.

3 Cited by Dalton 1971: ix.

4 Expression of E.H. Carr cited by Dalton (1971: xiii).

5 The symbolic nature of social existence is underscored by Voegelin in his project of 'A New Science of Politics.' He states: 'Human society is not merely a fact, or an event, in the external world to be studied by an observer like a natural phenomenon. Though it has externality as one of its important components, it is as a whole, a little world, a cosmion, illuminated with meaning from within by the human beings who continuously create and bear it as the mode and condition of their self-realization. It is illuminated through an elaborate symbolism in various degrees of compactness and differentiation – from rite, through myth, to theory, and this symbolism illuminates it with meaning insofar as the symbols made the internal structure of such a cosmion, the relations between its members and groups of members, as well as its existence as a whole, transparent for the mystery of human existence. The full self-illumination of society through symbols is an integral part of social reality, and one may even say its essential part, for through such symbolization the members of a society experience it as more than an accident or convenience; they experience it as of their human essence' (Voegelin 1969: 27).

6 See Polanyi 1971b. See also Bücher 1968.

7 For summary of Cassirer's theory see Cassirer 1944.

8 This expression is borrowed from Voegelin. When describing the tension immanent in human existence he accentuates its in-between structure. The vocabulary of the established model of social science, for obvious reasons, is inadequate for the conceptualization of this theme. In fact, Voegelin's restorative endeavor entails criteria of cognition and, therefore, of language which seem striking to those overconformed to the prevailing model of political science. For a characterization of Voegelin's model of political science, see Sandoz, 1972. On the notion of 'in-between,' see Voegelin 1970 and 1974.

9 See, on this point, Wrong 1961.

10 On this point see Weisskopf 1957 and 1971.

11 Cited in Hicks (1969: 123).

12 Sir John Hicks writes: 'Labor ... is not ... "work." Each of the classes of people whose activities we have been surveying has its work. The peasant has his work, the administrator has his work, the merchant has his work, even the landlord, so long as he retains a positive function, has his work. The characteristic of the laborer, or worker in the narrower sense ... is that he works for someone else. He is (let us not be afraid to say) a servant.' And Hicks adds:

'The Mercantile Economy has never been able to dispense with servants' (Hicks, 1969: 122).
13 Cited in Galbraith (1958: 264).
14 Schon's networks and network roles specifically illustrate such an approach. See Schon 1971.

References

Arendt, H. 1958. *The Human Condition*. Garden City, NY: Doubleday
Argyris, C. 1973a. 'Some Limits of Rational Man Organizational Theory,' *Public Administration Review*, May/June
– 1973b. *On Organizations of the Future*. Beverly Hills, Calif.: Sage Publications
Berger, P. and T. Luckman. 1967. *The Social Contruction of Reality*. Garden City, Doubleday
Blumer, H. 1962. 'Society as Symbolic Interaction,' in A.M. Rose ed, *Human Behavior and Social Processes*. Boston: Houghton Mifflin
Boguslaw, R. 1965. *The New Utopians*. New Jersey: Prentice-Hall
Bücher, Carl. 1968. *Industrial Evolution*. New York, Augustus M. Keller Publishers
Buckley, W. 1972. 'A Systems Approach to Epistemology,' in G.P. Klir ed, *Trends in General Systems Theory*. New York: Wiley-Interscience, a division of John Wiley and Sons
Cassirer, Ernst. 1944. *An Essay on Man*. New Haven, Conn.: Yale University Press
Churchman, C.W. 1972. *Designs of Inquiring Systems*. New York: Basic Books
– 1979. *The Systems Approach and its Enemies*. New York: Basic Books
Dalton, G. ed. 1971. *Primitive, Archaic, and Modern Economies: Essays of Karl Polanyi*. Boston: Beacon Press
Galbraith, J.K. 1958. *The Affluent Society*. New York: New American Library
– 1973. *Economics and the Public Purpose*. Boston: Houghton Mifflin
Gross, B. 1973. 'An Organized Society?' *Public Administration Review*, July/August
Gurvitch, G. 1971. *The Social Frameworks of Knowledge*. New York: Harper and Row
Habermas. Jürgen. 1970. *Towards a Rational Society*. Boston: Beacon Press
– 1971. *Knowledge and Human Interests*. Boston:Beacon Press
Hicks, J. 1969. *A Theory of Economics*. New York: Oxford University Press
Husserl, Edmund. 1967. 'The Thesis of the Natural Standpoint and Its Suspension,' in J. Kocklemans ed, *Phenomenology*. Garden City, Doubleday.
Mannheim, K. 1940. *Man and Society in an Age of Reconstruction*. New York: Harcourt, Brace and World
March, J. and Simon H. 1958. *Organizations*. New York: John Wiley and Sons
Moreno, J. 1934. *Who Shall Survive?* Washington, DC: Nervous and Mental Disease Publishing Co.
Neisser, U. 1967. *Cognitive Psychology*. New York: Appleton-Century-Crofts

Perrow, C. 1972. *Complex Organizations: A Critical Essay*. Glenview, Ill.: Scott, Foresman

Pfiffner, J. and F. Sherwood. 1969. *Administrative Organization*. Englewood Cliffs, Prentice-Hall

Pieper, J. 1963. *Leisure: The Basis of Culture*. New York: New American Library

Polanyi, K. 1971a. *The Great Transformation*. Boston: Beacon Press

– 1971b. 'Societies and Economic Systems,' in G. Dalton 1971.

Sandoz, E. 1972. 'The Philosophical Science of Politics,' in G. Graham and G. Carey eds, *The Post-Behavioral Era: Perspectives of Political Science*. New York: David McKay

Sartori, G. 1970. 'Concept Misformation in Comparative Politics,' *The American Political Science Review*, December

Schon, D. 1971. *Beyond the Stable State*. New York: Random House

Simon, H. 1965. *Administrative Behavior*. New York: The Free Press

– 1973. 'Organizational Man: Rational or Self-actualizing,' *Public Administration Review*, July/August

Strauss, L. 1972. 'Political Philosophy and the Crisis of Our Time,' in G. Graham and G. Carey eds, *The Post-Behavioral Era*. New York: David McKay

Voegelin, Eric 1964. *Order and History, Vol. II: The World of the Polis*. Baton Rouge: Louisiana State University Press

– 1969. *The New Science of Politics*. Chicago: University of Chicago Press

– 1970. 'Equivalences of Experiences and Symbolization in History,' *In Eternita E Storia*. Florence: Vallechi Editore

–1974. 'Reason: The Classical Experience,' *The Southern Review*, spring

Weisskopf, W.A. 1957. *The Psychology of Economics*. London: Routledge and Kegan Paul

– 1971. *Alienation and Economics*. New York: E.P. Dutton

Weizenbaum, Joseph. 1976. *Computer Power and Human Reason*. San Francisco: W.H. Freeman

Willey, B. 1953. *The Seventeenth Century Background*. Garden City, Doubleday

Wrong, D. 1961. 'The Oversocialized Conception of Man in Modern Sociology,' *American Sociological Review*, April

CHAPTER 7

Notes

1 See for instance, Monsen and Downs 1971 and Tullock 1972.
2 For a more complex view of unidimensionalization, see Marcuse 1966. On this process in historical perspective, see Halmos 1953.

3 On this point see Polanyi 1971.
4 See Roshwald 1973.
5 See especially Goffman 1961. See also Paul Goodman 1960.
6 See Schon 1971.
7 See Steele 1973.
8 On Alinsky see Norton 1972. See also Kotler 1969.

References

Goffman, E. 1961. *Asylums*. Garden City, NY: Doubleday
Goodman, Paul 1960. *Growing Up Absurd*. New York: Random House
Goodman, Paul and Percival. 1960. *Communitas: Means of Livelihood and Ways of life*. New York: Random House
Gottschalk, S.S. 1973. 'The Community-Based Welfare System: An Alternative to Public Values,' *The Journal of Applied Behavioral Science* 9: 213
Gross, B. 1973. 'An Organized Society?' *Public Administration Review*, July/August
Halmos, P. 1953. *Solitude and Privacy*. New York: Philosophical Library
Kotler, M. 1969. *Neighborhood Government*. Indianapolis: Bobbs Merrill
Marcuse, H. 1966. *One-Dimensional Man*. Boston: Beacon Press
McWhinney, W. 1973. 'Phenomenarchy: A Suggestion for Social Redesign,' *The Journal of Applied Behavioral Science* 9: 213
Monson, R.J. and A. Downs. 1971. 'Public Goods and Private Goods,' *Public Interest*, spring
Moreno, J. 1934. *Who Shall Survive?* Washington, DC: Nervous and Mental Disease Publishing Co.
Norton, E. 1972. 'Playboy Interview: Saul Alinsky,' *Playboy*, March
Polanyi, Karl. 1971. *The Great Transformation*. Boston: Beacon Press
Query, J.M.N. 1973. 'Total Push and the Open total Institution: The Factory Hospital,' *The Journal of Applied Behavioral Science* 9: 213
Roshwald, M. 1973. 'Order and Overorganization in America,' *The British Journal of Sociology* 24
Rush, H.M.F. 1969. *Behavioral Science: Concepts and Management Application*. New York: National Industrial Conference Board
Sarason, S.B. 1974. *The Psychological Sense of Community*. San Francisco: Jossey-Bass Publishers
Schon, D. 1971. *Beyond the Stable State*. New York: Random House
Shapiro, H.R. 1975. *The Bureaucratic State*. Brooklyn: Samizdat Press
Slater, P. 1971. *The Pursuit of Loneliness: American Culture at the Breaking Point*. Boston: Beacon Press
Steele, F. 1973. *Physical Setting and Organizational Development*. Reading, Mass.: Addison-Wesley Publishing

Thompson, V.A. 1966. *Modern Organization: A General Theory*. New York: Alfred A. Knopf
Tullock, G. 1972. 'Economic Imperialism,' in J.M. Buchanan and R.D. Tollinson eds, *Theory of Public Choice*. Ann Arbor: University of Michigan Press

CHAPTER 8

Notes

1 On the concept of 'objective possibility' see Ramos 1970.
2 On the notion of design in a broad sense see Pye 1968, 1969 and Thompson 1961.
3 On the technological dimension of social systems and its design implications see, for instance, Woodward 1965, Lawrence and Lorsch 1969, Perrow 1965, 1970, 1972, Burns and Stalker 1961, Thompson 1967, Davis and Engelstad 1966, Van Beinum 1968, Miller and Rice 1967, Emery 1969, Davis and Taylor 1972, and Davis and Cherns 1975.
4 I am deliberately avoiding the use of the controversial notion of 'optimal size.' On this question see Alonso 1971 and Richardson 1973. I am indebted to Helio Viana for calling my attention to this controversy.
5 On size of organizations see Schumacher 1973: 59–70. This statement focuses upon the intensity of face-to-face interpersonal relationships. Certainly today's technology of communications can integrate people in a community of interests regardless of great physical distances separating them. Taking into account this circumstance, Melvin M. Webber proposes the concept of 'non-place communities' (Webber 1967).
6 On Habermas and knowledge interests see the first chapter of this book.
7 A vivid assessment of the industrial revolution in England and of its economic, social, and architectural outcomes is provided by John Ruskin in his books *Unto This Last, Sesame and Lilies, The Ethics of Dust, The Crown of Wild Olive, Time and Tide, Fors Clavigera, Munera Pulversi*, and *The Stones of Venice*. The best edition of Ruskin's works is organized by Cook and Wedderburn (1903). Another kind of important assessment of the industrial revolution, emphasizing the social, cultural, and psychological outcomes of the widespread use of money is Simmel (1978).
8 See for instance Freedman 1975, Barash 1977, Wilson 1977, Sahlins 1976.
9 This is, however, an overdeterministic statement. Eventually man and even animals can transcend the distortive character of space. On this subject see, for instance, Frankl 1968 and Bettelheim 1958.

10 See H.M. Proshansky et al. 1970.

11 There are a few exceptions. See Lee 1968 and Waldo 1970.

12 On the philosophical approach to time see Fraser et al. 1972. Time and space are significant referents of Innis's historical studies. See Innis 1972.

13 On the notion of *apeiron* see Kahn 1960 and Seligman 1962.

14 See, for instance, Gordon 1973.

15 On this subject see Linder 1970.

16 See de Grazia 1964: 82.

17 See Cox 1970.

References

Alonso, W. 1971. 'The Economics of Urban Size,' *Papers of the Regional Science Association* 26

Ashby, W. Ross. 1968. 'Variety, Constraint, and the Law of Requisite Variety,' in W. Buckley ed, *Modern Systems Research for the Behavioral Scientist.* Chicago: Aldine Publishing

Barash, D.P. 1977. *Sociobiology and Behavior.* New York: Elsevier North-Holland

Bergson, Henri. 1956. *The Two Sources of Morality and Religion.* Garden City, NY: Doubleday

Bettelheim, H. 1958. 'Individual and Mass Behavior in Extreme Situations,' in E. Maccoby, et al. eds, *Readings in Social Psychology.* New York: Holt, Rinehart and Winston

Bierman, A.K. 1973. *The Philosophy of Urban Existence.* Athens, Ohio: Ohio University Press

Burns, T. and Stalker, G.M. 1961. *The Management of Innovation.* London: Tavistock

Cox, H. 1970. *The Feast of Fools.* New York: Harper & Row

Dahl, R. 1975. *After the Revolution?* New Haven, Conn.: Yale University Press

Dahl, R. and Tufte, E.R. 1973. *Size and Democracy.* Stanford, Calif.: Stanford University Press

Davis, L.E. and Cherns, A.B. eds. 1975. *The Quality of Working Life.* London: The Free Press, Collier Macmillan

Davis, L.E. and Engelstad, P.H. 1966. *Unit Operations in Socio-technical Systems: Analysis and Design.* London: Tavistock Institute of Human Relations. Doc. no. T894, October

Davis, L.E. and Taylor, J.C. 1972. *Design of Jobs.* Penguin Books

de Grazia, Sebastian. 1964. *Of Time, Work, and Leisure.* Garden City, NY: Doubleday

Emery, F.E. ed. 1969. *Systems Thinking.* Middlesex, England: Penguin Books

Follet, M.P. 1973. *Dynamic Administration.* Collected Papers of Mary Parker Follet. Edited by E.M. Fox and L. Urwick. London: Pitman Publishing

Frankl, Viktor 1968. *Man's Search for Meaning.* New York: Washington Square Press

Fraser, J.T., Haber, F.C., and Miller, G.H. eds. 1972. *The Study of Time.* Berlin, Heidelberg; Springer-Verlag

Freedman, J.L. 1975. *Crowding and Behavior.* New York: Viking Press

Goethe, J.W. von. 1949. *Goethe's Autobiography.* Washington, DC: Public Affairs Press

Gordon, W.J.J. 1973. *Synectics.* New York: Collier Books

Gunnel, J.G. 1968. *Political Philosophy and Time.* Middletown, Conn.: Wesleyan University Press

Gurvitch, G. 1955. *Déterminismes sociaux et liberté humaine. Paris: Presses Universitaires de France*

– 1964. *The Spectrum of Social Time.* Dordrecht: D. Reidel

– 1971. *The Social Frameworks of Knowledge.* New York: Harper & Row

Hall, E.T. 1959. *The Silent Language.* Greenwich, Conn.: Fawcett Publications

– 1966. *The Hidden Dimension.* Garden City, NY: Doubleday

Hesse, H. 1973. *Autobiographical Writings.* New York: Farrar, Straus and Gicroux

Innis, H.A. 1972. *Empire and Communications.* Toronto: University of Toronto Press

Jung, C.G. 1963. *Memories, Dreams, Reflections.* New York: Vintage Books

Kahn, C.H. 1960. *Anaximander and the Origins of Greek Cosmology.* New York: Columbia University Press

Kohr, L. 1978. *The Overdeveloped Nations.* New York: Schock Books

– 1978. *The Breakdown of Nations.* New York: E.P. Dutton

Kierkegaard, Søren 1962. *The Present Age.* New York: Harper & Row

Laing, R.D. 1967. *The Politics of Experience.* New York: Ballantine Books

Lawrence, P.R. and Lorsch, J.W. 1969. *Developing Organizations: Diagnosis and Action.* Reading, Mass.: Addison-Wesley Publishing

Lee, Hahn-Ben. 1968. *Korea: Time, Change and Administration.* Honolulu: East-West Center Press

Linder, S.B. 1970. *The Harried Leisure Class.* New York: Columbia University Press

Miller, E.J. and Rice, A.K. 1967. *Systems of Organization.* London: Tavistock Publications

Munthe, Axel. 1956. *The Story of San Michele.* New York: E.P. Dutton

Perrow, C. 1965. 'Hospitals: Technology, Goals, and Structure,' in J. March ed, *Handbook of Organizations.* Chicago: Rand McNally

– 1970. *Organizational Analysis. A Sociological View*. Belmont, Calif. in Wadsworth Publishing
– 1972. *Complex Organizations*. Glenview, Ill.: Scott, Foresman
Potter, D.M. 1954. *People of Plenty*. Chicago: University of Chicago Press
Progoff, Ira. 1973. *The Symbolic and the Real*. New York: McGraw-Hill
Proshansky, H.M., Ittelson, W.H., and Rivlin, L.G. eds. 1970. *Environmental Psychology: Man and His Physical Setting*. New York: Holt, Rinehart and Winston
Pye, D. 1968. *The Nature and Art of Workmanship*. Cambridge: Cambridge University Press
– 1969. *The Nature of Design*. New York: Reinhold Book Corp.
Ramos, A.G. 1970. 'Modernization: Towards a Possibility Model,' in W.A. Beling and G.O. Totten eds, *Developing Nations: Quest for a Model*. New York: Von Nostrand Reinhold
Richardson, H.W. 1973. *The Economics of Urban Size*. Lexington, Mass.: Lexington Books, D.C.Heath
Ruskin, John. 1903. *The Works of John Ruskin*. Edited by E.T. Cook and A. Wedderburn. London: George Allen
Sahlins, M. 1976. *The Use and Abuse of Biology*. Ann Arbor, Mich.: University of Michigan Press
Schumacher, E.F. 1973. *Small Is Beautiful*. New York: Harper & Row
Seligman, P.C. 1962. *The Apeiron of Anaximander*. London: Athlone Press
Simmel, G. 1950. *The Sociology of Georg Simmel*. Translated, edited, and with an introduction by K.H. Wolff. New York: The Free Press
– 1978. *The Philosophy of Money*. Boston: Routledge & Kegan Paul
Skolimowski, H. 1975. 'Teilhard, Soleri and Evolution,' *The Teilhard Review* 10, no. 3
Sommer, R. 1969. *Personal Space*. Englewood Cliffs, NJ: Prentice-Hall
– 1972. *Design Awareness*. San Francisco: Reinhart Press
Steele, F.I. 1973. *Physical settings and Organization Development*. Reading, Mass.: Addison-Wesley Publishing
Thompson, D'Arcy W. 1961. *On Growth and Form*. Cambridge: Cambridge University Press
Thompson, J. 1967. *Organizations in Action*. New York: McGraw-Hill
Tuan, Yi-Fu. 1974. *Topophilia*. Englewood Cliffs, NJ: Prentice-Hall
Van Beinum, H.J.J. 1968. *The Design of the New Radial Tyre Factory as an Open Socio-technical System*. London: Tavistock Institute of Human Relations, HRC 150, 28 October
Waldo, D. ed. 1970. *Temporal Dimensions of Development Administration*. Durham, NC: Duke University Press

Webber, M.M. et al. 1967. *Explorations into Urban Structure*. Philadelphia: University of Pennsylvania Press

Wilson, E.O. 1977. *Sociobiology: The New Synthesis*. Cambridge, Mass.: The Belknap Press of Harvard University Press

Woodward, J. 1965. *Industrial Organizations: Theory and Practice*. London: Oxford University Press

CHAPTER 9

Notes

1 For instance, there is a para-economic overtone in the works and thinking of individuals like Kenneth Boulding, Barry Commoner, René Dubos, Gunnar Myrdal, C.B. Macpherson, John Gardner, Ralph Nader, and Hazel and Carter Henderson. What is most characteristic of these individuals is their value-commitment, their confrontive posture toward prevailing types of organizational arrangements. From the normative perspective of the paradigm, the para-economic consultant would be selective in accepting assignments because he is willing to lend his expertise only to promotional attempts at creating and implementing non-market designs of personal and collective living, or to existing economies where he sees a proclivity toward changes which better enable them to meet genuine individual and public needs.

  Thus, the para-economist should not be confused with what Alvin Toffler calls the adhocrat. Toffler defines *adhocracy* as a task force which helps organizations to accomplish their goals without systematically questioning the nature of such goals. Toffler's *adhocracy* is an outcome of a type of reactive thinking which considers the present all-inclusive market system as a given and, therefore, seeks to legitimize the changes engendered by its intrinsic dynamics. He specifically conceives it as a tool for enhancing the existing economies' 'cope-ability' (Toffler 1970: 357).

  In this sense there is in the professional activity of the adhocrat or the common consultant no delimitative intentionality. In opposition to this orientation, para-economy is conceived as a category of confrontive and delimitative thinking. Thus the para-economic consultant is determined to work only for changes which are meaningful from the standpoint of his personal paradigm of the good order of human and social affairs.

  There are today few persons who could be classified as para-economic activists. However, a para-economic posture has increasingly become a salient dimension of first-class consultants in this country. For instance, A.K. Bierman comes quite close to what can be envisioned as a para-economic change agent. He has participated in some neighborhood programs in San Francisco according to what deserves to be called a para-economic strategy. First of all, Bierman's action reflects his views of what a city should be, as articulated in his book *The Philosophy of Urban Existence*. Like Milton Kotler in his proposals for neighborhood governments, Bierman realizes that the policies followed by county authorities to develop the arts usually reinforce the 'imperialism' of downtown in relation to the community at large. He points out that the 'art center museum mentality that seems to hypnotize New York, Los Angeles, and Washington' actually helps 'to preserve the value of downtown real estate' and to 'pipe suburbanites back into being urbanites, if only for a few ticketspree nights of the year' (Bierman 1973: 183). The neighborhood arts program, which Bierman helped to establish, resisted such a centralizing policy. The leadership of the program was able to persuade local foundations and the mayor and supervisors to contribute to setting up funds amounting to several million dollars. The success of this program has led Bierman to believe that the idea of neighborhood arts is strong enough in San Francisco to serve as a viable alternative to the traditional model of central arts.

The systematic implications associated here with the category of para-economy may also pervade Donald Schon's endeavors. In *Beyond the Stable State* Schon suggests that at present the U.S. Government is deprived of institutional capabilities to meet the needs of our complex society. One main reason for this institutional handicap is the centralized policy-making system on the basis of which the government deals with administrative agencies at state and local levels as if it were their preceptor. Innovations at those levels are stifled by this overcentralized policy-making model. Schon acknowledges the need to overcome the 'dynamic conservatism' of the governmental centers of policy and deems it necessary to leave more room for the decentralized initiation and implementation of public policies. In order to transform the present government into a public learning system, he suggests the 'design, development and management of networks' (Schon 1971: 190) which will enable the central government 'to function as facilitator of society's learning, rather than as society trainer' (178). Network management, as he conceives it, is obviously a confrontive approach to social systems design, and supposedly the *Organization for Social and Technological Innovation* (OSTI) over which Schon presides at Cambridge, Massachusetts, is to some extent an illustration of a para-economic agency. Moreover, in other books Schon has also developed a methodology for innovation in general and in technological terms. These may be important subsidiary elements for the creation and implementation of 'convivial' and related designs similar to those proposed by Ivan Illich, E.F. Schumacher and Victor Papanek, which are advocated as necessary counters to the all-social inclusiveness of the present industrial market system.

2 See on this Boulding 1973, Boulding and M. Pfaff 1972, and Boulding, M. Pfaff, and A. Pfaff 1973.

3 Expanding the theoretical framework presented in this book, George K. Najjar has focused upon budgeting as a tool of economic development (Najjar 1978).

4 On this see Tribe 1972.

5 See Churchman 1971; Tribe 1971, 1973, 1976; Dolbeare 1975; Kramer 1975.

6 In his remarkable book *Politics and Markets* Lindblom points out several distortive social and political outcomes of contemporary market systems, but as in 1953 he never systematically addressed himself to the questions implied by delimitation.

7 See for instance Mishan 1977, Ul Haq 1976, Seers 1977, Frank 1972, Holsti 1975, Streeten 1977, Morinson 1974.

8 On the variety of these initatives see Carter Henderson 1977, 1978; Hazel Henderson 1978; Stravianos 1976; the special issue of *Journal of Applied Behavioral Science* 9, 1973; Berger and Neuhaus 1977; Gershuny 1978.

9 See Yankelovich 1978.

10 See H. Henderson (1978: 390 and 1978a). See also her forthcoming book *The Politics of Reconceptualization*.

11 See Georgescu-Roegen (1976: 20).

12 See Ramos 1970.

References

Berger, P.L. and Neuhaus, R.J. 1977. *To Empower People*. Washington, DC: American Enterprise Institute for Public Policy Research

Bierman, A.K. 1973. *The Philosophy of Urban Existence*. Athens, Ohio: Ohio University Press

Boulding, K.E. 1973. *The Economy of Love and Fear*. Belmont, Calif.: Wadsworth Publishing

Boulding, K.E. and Pfaff, M. eds. 1972. *Redistribution to the Rich and the Poor: The Grant Economics of Income Distribution*. Belmont, Calif.: Wadsworth Publishing

Boulding, K.E., Pfaff, M. and Pfaff, A. eds. 1973. *Transfers in an Urbanized Economy*. Belmont, Calif.: Wadsworth Publishing

Brown, H. 1978. *The Human Future Revisited*. New York: W.W. Norton

Burns, Scott. 1975. *The Household Economy*. Boston: Beacon Press

Churchman, C.W. 1971. *The Design of Inquiring Systems*. New York: Basic Books

De Grazia, Sebastian. 1964. *Of Time, Work and Leisure*. Garden City, NY: Doubleday

Dahl, R.A. and Lindblom, C.E. 1963. *Politics, Economics and Welfare*. New York: Harper & Row

Daly, H.D. 1977. *Steady-State Economy*. San Francisco: W.H. Freeman

Dolbeare, K.M. 1975. 'Public Policy Analysis and the Coming for the Soul of the Postbehavioral Revolution,' in P. Green and S. Levinson, eds, *Power and Community: Dissenting Essays in Political Science*. New York: Vintage Books

Ferguson, Marilyn 1980. *The Aquarian Conspiracy*. New York: St. Martin Press

Frank, A.G. 1972. 'Sociology of Development and Underdevelopment of Sociology,' in J.D. Crockroft et al. eds, *Dependence and Underdevelopment*. Garden City, NY: Doubleday

Georgescu-Roegen, N. 1976. *Energy and Economic Myths*. New York: Pergamon Press

– 1976a. *The Entropy Law and the Economic Process*. Cambridge, Mass.: Harvard University Press

Gershuny, J. 1978. *After Industrial Society? The Emerging Self-Service Economy*. Atlantic Highlands, NJ: Humanities Press

Henderson, C. 1977. 'Living the Simple Life,' *Human Resource Management* 16, no. 3

– 1978. *New Age Enterprise*. Brochure. Princeton, NJ: Princeton Center for Alternative Futures

Henderson, H. 1978. *Creating Alternative Futures*. New York: Berkly Windhover Books

– 1978a. 'Risk, Uncertainty and Economic Futures,' *Best's Review*, May

Holsti, K.I. 1975. 'Underdevelopment and the "Gap" Theory of International Conflict,' *American Political Science Review* 63, no. 3

Kramer, F.A. 1975. 'Policy Analysis as Ideology,' *Public Administration Review*, September/October

Lindblom, C.E. 1977. *Politics and Markets*. New York: Basic Books

Meadows, D.L. ed. 1977. *Alternatives to Growth (I)*. Cambridge, Mass.: Ballinger Publishing

Mishan, E.J. 1977. *The Economic Growth Debate*. London: George Allen & Unwin

Morinson, E.E. 1974. *From Know-How to Nowhere*. New York: Basic Books

Najjar, G.K. 1978. 'Social Systems Delimitation and Allocative Mechanisms,' *Administration and Society* 9, no. 4

Ophuls, W. 1977. *Ecology and the Politics of Scarcity*. San Francisco: W.H. Freeman

Ramos, A.G. 1970. 'Modernization: Towards the Possibility Model,' in W.A. Beling and G.O. Totten eds, *Developing Nations, Quest For a Model*. New York: Van Nostrand Reinhold

Schon, D. 1971. *Beyond the Stable State*. New York: Random House

Seers, D. 1977. 'The New Meaning of Development,' *International Development Review* 19, no. 3

Stavrianos, L.S. 1976. *The Promise of the Coming Dark Age*. San Francisco: W.H. Freeman

Streeten, P. 1977. 'Changing Perceptions of Development,' *Finance and Development* 14, no. 3

Toffler, A. 1970. *Future Shock*. New York: Random House

Tribe, L.A. 1971. 'Legal Framework for the Assessment and Control of Technology,' *Minerva* 73, March

– 1972. 'Policy Science: Analysis or Ideology,' *Philosophy & Public Affairs* 2, no. 1 fall

– 1973. 'Technology and the Fourth Discontinuity: The Limits of Instrumental Rationality,' *Southern California Law Review* 46

Tribe, L.A. et al. eds. 1976. *When Values Conflict*. Cambridge, Mass.: Ballinger Publishing

Ul Haq, M. 1976. 'Towards a Just Society,' *International Development Review* 18, no. 4

Yankelovich, D. 1978. 'The New Psychological Contracts at Work,' *Psychology Today*, May

CHAPTER 10

Notes

1 It will be obvious to those familiar with Whitehead's theory that this analysis is generally influenced by his thought. I must warn the reader, however, that my use of the word 'endurance' may not be fully consistent with Whitehead's. My substantiation of an enlarged notion of endurance cannot be developed within the limits of these concluding remarks.

2 See my article 'Endurance and Fluidity: A Reply,' *Administration and Society*, February 1977.

3 In order to see how much of the old classical thought is relevant to contemporary attempts at reformulating social science, see Sibley 1973.

References

Alexander, C. 1974. *Notes on the Synthesis of Form*. Cambridge, Mass.: Harvard University Press

Georgescu-Roegen, N. 1973. 'The Entropy Law and the Economic Problem,' in
    H.E. Daly, *Toward a Steady State Economy*. San Francisco: W.H. Freeman
Likert, R. 1963. 'Trends toward a World-Wide Theory of Management,' cios 13
Sibley, M.Q. 1973. 'The Relevance of Classical Political Theory for Economy,
    Technology & Ecology,' *Alternatives* 2, no. 2
Whitehead, A.N. 1967. *Science and the Modern World*. New York: The Free Press

Index